Contents

v

Participants

E. Boncinelli   DIBIT, Scientific Institute San Raffaele, Via Olgettina 58, Milan, Italy

F. Bonhoeffer   MPI für Entwicklungsbiologie, Abt. Physikal. Biologie, Spemannstr. 35/1, 72076 Tübingen, Germany

V. Broccoli (*Bursar*)   TIGEM Institute Dibit HS Raffaele, Via Olgettina 58, I-20132 Milan, Italy

A. B. Butler   Krasnow Institute for Advanced Study and Department of Psychology, George Mason University, MSN 2A1, Fairfax, VA 22030, USA

S. E. Evans   Department of Anatomy and Developmental Biology, University College London, Gower Street, London WC1E 6BT, UK

A. M. Goffinet   Neurobiology Unit, University of Namur Medical School, 61 rue de Bruxelles, B5000 Namur, Belgium

K. Herrup   Department of Neuroscience and University Alzheimer Research Center of Cleveland, Case Western Reserve University, 10900 Euclid Avenue Cleveland, OH 44120, USA

W. Hodos   Department of Psychology, University of Maryland, College Park, MD 20742-4411, USA

S. Hunt   Department of Anatomy and Developmental Biology, Medawar Building, University College London, Gower Street, London WC1E 6BT, UK

J. H. Kaas   Department of Psychology, Vanderbilt University, 301 Wilson Hall, Nashville, TN 37240, USA

H. J. Karten (*Chair*)   Department of Neuroscience, University of California at San Diego, La Jolla, CA 92093-0608, USA

L. A. Krubitzer   Center for Neuroscience and Department of Psychology, 1544 Newton Court, University of California, Davis, CA 95616, USA

P. R. Levitt   Department of Neurobiology, University of Pittsburgh School of Medicine, E1440 Biomed Science Tower, Pittsburgh, PA 15261, USA

A. Lumsden   Department of Developmental Neurobiology, King's College London, Hodgkin House, Guy's Campus, Guy's Hospital, London SE1 9RT, UK

Z. Molnár*   Institut de Biologie Cellulaire et de Morphologie, Université de Lausanne, Rue du Bugnon 9, 1005 Lausanne, Switzerland

D. D. M. O'Leary   Laboratory of Molecular Neurobiology, The Salk Institute, 10010 North Torrey Pines Road, La Jolla, CA 92037, USA

N. Papalopulu   Wellcome/CRC Institute, Tennis Court Road, Cambridge CB2 1QR, UK

J. G. Parnavelas   Department of Anatomy and Developmental Biology, University College London, Gower Street, London WC1E 6BT, UK

J. Pettigrew   Vision Touch and Hearing Research Centre, University of Queensland, Brisbane, QLD 4072, Australia

L. Puelles   Dpto. Ciencias Morfologicas, Facultad de Medicina, Universidad de Murcia, Campus de Espinardo, 30100 Espinardo, Murcia, España

D. Purves   Department of Neurobiology, Box 3209, Duke University Medical Center, 101-I Bryan Research Building, Durham, NC 27710, USA

P. Rakic   Section of Neurobiology, Yale University School of Medicine, New Haven, CT 06510, USA

A. J. Reiner   Department of Anatomy and Neurobiology, College of Medicine, University of Tennessee, 855 Monroe Avenue, Memphis, TN 38163, USA

---

*Current address: Department of Human Anatomy and Genetics, University of Oxford, South Parks Road, Oxford OX1 3QX, UK

J. L. R. Rubenstein   Nina Ireland Laboratory of Developmental Neurobiology, Center for Neurobiology and Psychiatry, Department of Psychiatry and Programs in Neuroscience, Developmental Biology and Biomedical Sciences, University of California at San Francisco, 401 Parnassus Avenue, San Francisco, CA 94143, USA

E. Welker   Institut de Biologie Cellulaire et de Morphologie, Université de Lausanne, Rue du Bugnon 9, 1005 Lausanne, Switzerland

L. Wolpert   Department of Anatomy and Developmental Biology, University College London, Gower Street, London WC1E 6BT, UK

What is evolutionary developmental biology?

L. Wolpert

Department of Anatomy and Developmental Biology, University College London, London WC1E 6BT, UK

Abstract. All changes in animal form and function during evolution are due to changes in their DNA. Such changes determine which proteins are made, and where and when, during embryonic development. These proteins thus control the behaviour of the cells of the embryo. In evolution, changes in organs usually involve modification of the development of existing structures—tinkering with what is already there. Good examples are the evolution of the jaws from the pharyngeal arches of jawless ancestors, and the incus and stapes of the middle ear from bones originally at the joint between upper and lower jaws. However, it is possible that new structures could develop, as has been suggested for the digits of the vertebrate limb, but the developmental mechanisms would still be similar. It is striking how conserved developmental mechanisms are in pattern formation, both with respect to the genes involved and the intercellular signals. For example, many systems use the same positional information but interpret it differently. One of the ways the developmental programmes have been changed is by gene duplication, which allows one of the two genes to diverge and take on new functions—*Hox* genes are an example. Another mechanism for change involves the relative growth rates of parts of a structure.

2000 Evolutionary developmental biology of the cerebral cortex. Wiley, Chichester (Novartis Foundation Symposium 228) p 1–14

It has been suggested that nothing in biology makes sense unless viewed in the light of evolution. Certainly it would be difficult to make sense of many aspects of development without an evolutionary perspective. Every structure has two histories: one that relates to how it developed, i.e. ontogeny; and the other its evolutionary history, i.e. phylogeny. Ontogeny does not recapitulate phylogeny as Haeckel once claimed, but embryos often pass through stages that their evolutionary ancestors passed through. For example, in vertebrate development despite different modes of early development, all vertebrate embryos develop to a similar phylotypic stage after which their development diverges. This shared phylotypic stage, which is the embryonic stage after neurulation and the formation of the somites, is probably a stage through which some distant

1

ancestor of the vertebrates passed. It has persisted ever since, to become a fundamental characteristic of the development of all vertebrates, whereas the stages before and after the phylotypic stage have evolved differently in different organisms.

Such changes are due to changes in the genes that control development. These control which proteins are made at the right time and place in the development of the embryo since it is proteins that determine how cells behave. One of the most important concepts in evolutionary developmental biology is that any developmental model for a structure must be able to account for the development of earlier forms in the ancestors.

Comparisons of embryos of related species has suggested an important generalization: the more general characteristics of a group of animals, that is those shared by all members of the group, appear earlier in evolution. In the vertebrates, a good example of a general characteristic would be the notochord, which is common to all vertebrates, and is also found in other chordate embryos. Paired appendages, such as limbs, which develop later, are special characters that are not found in other chordates, and that differ in form among different vertebrates. All vertebrate embryos pass through a related phylotypic stage, which then gives rise to the diverse forms of the different vertebrate classes. However, the development of the different vertebrate classes before the phylotypic stage is also highly divergent, because of their different modes of reproduction; some developmental features that precede the phylotypic stage are evolutionarily highly advanced, such as the formation of a trophoblast and inner cell mass by mammals.

Branchial arches

An embryo's development reflects the evolutionary history of its ancestors. Structures found at a particular embryonic stage have become modified during evolution into different forms in the different groups. In vertebrates, one good example of this is the evolution of the branchial arches and clefts that are present in all vertebrate embryos, including humans. These are not the relics of the gill arches and gill slits of an adult fish-like ancestor, but of structures that would have been present in the embryo of the fish-like ancestor. During evolution, the branchial arches have given rise both to the gills of primitive jawless fishes and, in a later modification, to jaws (Fig. 1). When the ancestor of land vertebrates left the sea, gills were no longer required but the embryonic structures that gave rise to them persisted. With time they became modified, and in mammals, including humans, they now give rise to different structures in the face and neck. The cleft between the first and second branchial arches provides the opening for the

Hypothetical ancestral jawless vertebrate

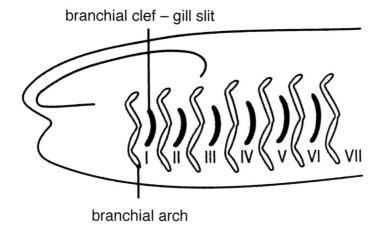

Jawed vertebrate

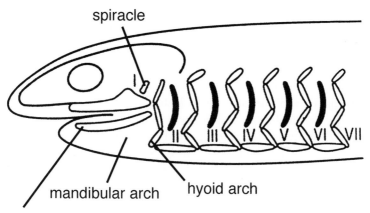

FIG. 1.   The ancestral jawless fish had a series of seven gill slits — branchial clefts — supported by cartilaginous or bony arches. Jaws developed from modification of the first arch (from Wolpert et al 1998).

Eustachian tube, and endodermal cells in the clefts give rise to a variety of glands, such as the thyroid and thymus (Fig. 2).

Evolution rarely generates a completely novel structure out of the blue. New anatomical features usually arise from modification of an existing structure. One

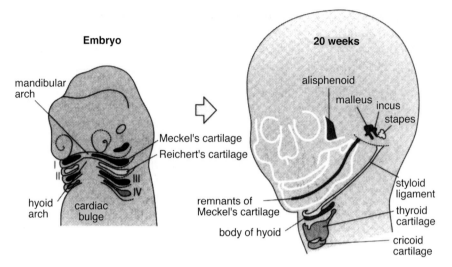

FIG. 2. Fate of branchial arch cartilage in humans. Cartilage in the branchial arches in the embryo give rise to elements that include the three auditory ossicles: the malleus and incus come from the first arch and the stapes from the second (from Wolpert et al 1998).

can therefore think of much of evolution as a 'tinkering' with existing structures, which gradually fashions something different. A nice example of a modification of an existing structure is provided by the evolution of the mammalian middle ear. This is made up of three bones that transmit sound from the eardrum (the tympanic membrane) to the inner ear. In the reptilian ancestors of mammals, the joint between the skull and the lower jaw was between the quadrate bone of the skull and the articular bone of the lower jaw, which were also involved in transmitting sound. During mammalian evolution, the lower jaw became just one bone, the dentary, with the articular no longer attached to the lower jaw. By changes in the development, the articular and the quadrate bones in mammals were modified into two bones, the malleus and the incus, whose function was now to transmit sound from the tympanic membrane to the inner ear. The skull bones of fish remain unfused and retain the segmental series of the gill arches.

Positional information

One of the ways that the embryo uses to make patterns and organs is based on positional information, that is the cells acquire a positional value related to boundary regions and then interpret this according to their genetic constitution and developmental history. Studies on regeneration of newt limbs and insect tibia show clearly that even adult cells can retain their positional values and generate new ones. One of the ways position can be specified during development is by a concentration gradient of a diffusible morphogen. This has several important implications for evolution. It means that a major change in development of the embryo comes from changes in interpretation of positional information, that is the cells' responses to signals. In fact there are a rather limited number of signalling molecules in most embryos—these include the transforming growth factor β (TGFβ) family, fibroblast growth factors (FGFs), sonic hedgehog, Wnts, Notch–delta, the ephrins and epidermal growth factors (EGFs). Evolution is both conservative and lazy, using the same signals again and again both within the same embryo and in other distantly related species; most of the key genes in vertebrate development are similar to those in *Drosophila*. Patterning using positional information allows for highly localized changes in the interpretation of position at particular sites. It is also a feature of development that the embryo at an early stage is broken up into largely independent 'modules' of a small size which are under separate genetic control. There is also good evidence that many structures make use of the same positional information but interpret it differently because of their developmental history. A classic case is that of the antenna and leg of *Drosophila*. A single mutation can convert an antenna into a leg and by making genetic mosaics it was shown that they use the same positional information but interpret it differently because of their developmental history—the antenna is in the anterior region of the body. Similar considerations apply to the fore- and hindlimbs of vertebrates. These differences in interpretation involve the *Hox* genes.

   Hox genes are members of the homeobox gene family, which is characterized by a short 180 base pair motif, the homeobox, which encodes a helix-turn-helix domain that is involved in transcriptional regulation. Two features characterize all known *Hox* genes: the individual genes are organized into one or more gene clusters or complexes, and the order of expression of individual genes along the anteroposterior axis is usually the same as their sequential order in the gene complex.

   Hox genes are key genes in the control of development and are expressed regionally along the anteroposterior axis of the embryo. The apparent universality of *Hox* genes, and certain other genes, in animal development has led to the concept of the zootype. This defines the pattern of expression of these key genes along the anteroposterior axis of the embryo, which is present in all animals.

   The role of the *Hox* genes is to specify positional identity in the embryo rather than the development of any specific structure. These positional values are interpreted differently in different embryos to influence how the cells in a region develop into, for example, segments and appendages. The *Hox* genes exert this influence by their action on the genes controlling the development of these structures. Changes in the downstream targets of the *Hox* genes can thus be a major source of change in evolution. In addition, changes in the pattern of *Hox* gene expression along the body can have important consequences. An example is a relatively minor modification of the body plan that has taken place within vertebrates. One easily distinguishable feature of pattern along the anteroposterior axis in vertebrates is the number and type of vertebrae in the main anatomical regions — cervical (neck), thoracic, lumbar, sacral and caudal. The number of vertebrae in a particular region varies considerably among the different vertebrate classes — mammals have seven cervical vertebrae, whereas birds can have between 13 and 15. How does this difference arise? A comparison between the mouse and the chick shows that the domains of *Hox* gene expression have shifted in parallel with the change in number of vertebrae. For example, the anterior boundary of *Hoxc6* expression in the mesoderm in mice and chicks is always at the boundary of the cervical and thoracic regions. Moreover, the *Hoxc6* expression boundary is also at the cervical–thoracic boundary in geese, which have three more cervical vertebrae than chicks, and in frogs, which only have three or four cervical vertebrae in all. The changes in the spatial expression of *Hoxc6* correlate with the number of cervical vertebrae. Other *Hox* genes are also involved in the patterning of the anteroposterior axis, and their boundaries also shift with a change in anatomy.

Thus a major feature of evolution relates to the downstream targets of the *Hox* genes. Unfortunately, these are largely unknown, but they are a major research area.

There is thus the conservation of some developmental mechanisms at the cellular and molecular level among distantly related organisms. The widespread use of the *Hox* gene complex and of the same few families of protein signalling molecules provide excellent examples of this. It seems that when a useful developmental mechanism evolved, it was used again and again. Bird wings and insect wings have some rather superficial similarities and have similar functions, yet are different in their structure. The insect wing is a double-layered epithelial structure, whereas the vertebrate limb develops mainly from a mesenchymal core surrounded by ectoderm. However, despite these great anatomical differences, there are striking similarities in the genes and signalling molecules involved in patterning insect legs, insect wings and vertebrate limbs. All these relationships suggest that, during evolution, a mechanism for patterning and setting up the axes of appendages appeared in some common ancestor of insects and vertebrates. Subsequently, the genes and signals involved acquired different downstream targets so that they could interact with different sets of genes, yet the same set of signals retain their organizing function in these different appendages. The individual genes involved in specifying the limb axes are probably more ancient than either insect or vertebrate limbs.

Gene duplication

A major general mechanism of evolutionary change has been gene duplication. Tandem duplication of a gene, which can occur by a variety of mechanisms during DNA replication, provides the embryo with an additional copy of the gene. This copy can diverge in its nucleotide sequence and acquire a new function and regulatory region, so changing its pattern of expression and downstream targets without depriving the organism of the function of the original gene. The process of gene duplication has been fundamental in the evolution of new proteins and new patterns of gene expression; it is clear, for example, that the different haemoglobins in humans have arisen as a result of gene duplication.

One of the clearest examples of the importance of gene duplication in developmental evolution is provided by the *Hox* gene complexes. Comparing the *Hox* genes of a variety of species, it is possible to reconstruct the way in which they are likely to have evolved from a simple set of six genes in a common ancestor of all species. Amphioxus, which is a vertebrate-like chordate, has many features of a primitive vertebrate: it possesses a dorsal hollow nerve cord, a notochord and segmental muscles that derive from somites. It has only one *Hox* gene cluster, and one can think of this cluster as most closely resembling the common ancestor of the four vertebrate *Hox* gene complexes — *Hoxa*, *Hoxb*, *Hoxc* and *Hoxd*. It is possible that both the vertebrate and *Drosophila Hox* complexes evolved from a simpler ancestral complex by gene duplication.

Limbs

The limbs of tetrapod vertebrates are special characters that develop after the phylotypic stage. Amphibians, reptiles, birds and mammals have limbs, whereas fish have fins. The limbs of the first land vertebrates evolved from the pelvic and pectoral fins of their fish-like ancestors. The basic limb pattern is highly conserved in both the fore- and hindlimbs of all tetrapods, although there are some differences both between fore- and hindlimbs, and between different vertebrates.

The fossil record suggests that the transition from fins to limbs occurred in the Devonian period, between 400 and 360 million years ago. The transition probably occurred when the fish ancestors of the tetrapod vertebrates living in shallow waters moved onto the land. The fins of Devonian lobe-finned fish the proximal skeletal elements corresponding to the humerus, radius and ulna of the tetrapod limb are present in the ancestral fish, but there are no structures corresponding to digits. How did digits evolve? Some insights have been obtained by examining the development of fins in a modern fish, the zebrafish. The fin buds of the zebrafish embryo are initially similar to tetrapod limb buds, but important differences soon arise during development. The proximal part of the fin bud gives rise to skeletal elements, which are homologous to the proximal skeletal elements of the tetrapod limb. There are four main proximal skeletal elements in a zebrafish fin which arise from the subdivision of a cartilaginous sheet. The essential difference between fin and limb development is in the distal skeletal elements. In the zebrafish fin bud, an ectodermal fin fold develops at the distal end of the bud and fine bony fin rays are formed within it. These rays have no relation to anything in the vertebrate limb.

If zebrafish fin development reflects that of the primitive ancestor, then tetrapod digits are novel structures, whose appearance is correlated with a new domain of *Hox* gene expression. However, they may have evolved from the distal recruitment of the same developmental mechanisms and processes that generate the radius and ulna. There are mechanisms in the limb for generating periodic cartilaginous structures such as digits. It is likely that such a mechanism was involved in the evolution of digits by an extension of the region in which the embryonic cartilaginous elements form, together with the establishment of a new pattern of *Hox* gene expression in the more distal region.

Growth and timing

Many of the changes that occur during evolution reflect changes in the relative dimensions of parts of the body. Growth can alter the proportions of the human baby after birth, as the head grows much less than the rest of the body. The variety of face shapes in the different breeds of dog, which are all members of the same species, also provides a good example of the effects of differential growth after birth. All dogs are born with rounded faces; some keep this shape but in others the nasal regions and jaws elongate during growth. The elongated face of the baboon is also the result of growth of this region after birth.

Because structures can grow at different rates, the overall shape of an organism can be changed substantially during evolution by heritable changes in the duration of growth that lead to an increase in the overall size of the organism. In the horse, for example, the central digit of the ancestral horse grew faster than the digits on either side, so that it ended up longer than the lateral digits.

Differences among species in the time at which developmental processes occur relative to one another can have dramatic effects on structures. For example, differences in the feet of salamanders reflects chances in timing of limb development; in an arboreal species the foot seems to have stopped growing at an earlier stage than in the terrestrial species. And in legless lizards and some snakes the absence of limbs is due to development being blocked at an early stage.

Further reading

Raff RA 1996 The shape of life: genes, development, and the evolution of animal form. University of Chicago Press, Chicago, IL

Wolpert L, Beddington R, Brockes J, Jessell T, Lawrence P, Meyerowitz E 1998 Principles of development. Oxford University Press, Oxford

DISCUSSION

Rakic: I enjoyed your presentation, but you didn't mention the importance of the nematode.

Wolpert: I'll tell you why I didn't mention nematode. In my opinion, studies of the nematode have not generally helped our understanding of the development of vertebrates, with the exception of insights into cell death, the netrins and signal transduction.

Rakic: As I will illustrate in my presentation, this may be quite significant. Furthermore, if you assumed that the roles of genes do not change in evolution, you would not be able to draw any conclusions concerning nematodes and humans. However, as you have said, genes are conserved, but their roles may be modified in different contexts. An example of this is the *sel2* gene, which was identified in the nematode and encodes a protein similar to Si28, which has been implicated in the early onset of Alzheimer's disease (Levitan & Greenwald 1995).

Wolpert: My position on the nematode is that it is peculiar, in the sense that specification of cell identity is on a cell-by-cell basis, whereas in *Drosophila* and in vertebrates it is on groups of cells. This is why the nematode doesn't tell us a great deal about vertebrates.

Herrup: I find it valuable for looking at vertebrates because, as you said, what is important is not so much the signal itself, but how the cells respond to the signal, and in *Caenorhabditis elegans*, you have to work on that problem at the level of the single cell. Therefore, it's a treat to see one cell doing what an entire cortex full of neurons are persuaded to do by their genes. However, I do agree that it is not useful for studying some of the more complex networks in *Drosophila*, for example.

Levitt: An example of conservation of signalling molecules occurs in the development of the *C. elegans* vulva. If the organization of epidermal growth factor (EGF)-like receptors is altered—and it is also possible to do this in vertebrates—the way the cell interprets the signal is changed, so that the cell develops into a different cell type. Maybe intracellular tinkering is what *C. elegans* does best.

Herrup: I would like to pursue the topic of digit development. What are the current theories as to how this occurs, and what can it tell us about how a small region at the end of a specialized structure can become an apparently novel morphological structure?

Wolpert: I wish I knew the answer. During the development of the proximal elements of the zebrafish, a sheet of cartilage breaks up into four elements. Therefore, the zebrafish has a mechanism to make repeated elements. Presumably, this is primitive and could have been used for making digits. Timing is an important issue in evolution because changes in timing can produce dramatic effects—if development continues for a longer period of time, then it may give rise to repeated structures at the ends of the digits. Conversely, if limb development stops early is reduced then this could give rise to loss of digits or even loss of limbs, as in legless lizards and snakes.

Karten: But there's much more to diversity than this, so the question becomes, what are the properties that confer these differences? We are finding that many organisms have common mechanisms, but this doesn't mean they're the same. Some of the issues concerning derivative gene families and gene duplication are beginning to give us hints about what underlies diversity and specialization, but how can we reconcile the constancies in evolution with the divergences that we observe? And can we specify the mechanisms for this?

Wolpert: The way I think about this is to consider the downstream targets. We don't understand how an antenna develops differently from a leg, and I can't think of an example of how downstream targets of a *Hox* gene control morphology.

Karten: This brings up another critical issue. We talk about high penetrance and the expression levels of particular genes. For instance, *Pax6* is expressed in the eyes of several animals, and it is also expressed in the olfactory placode. Are we confounding our search for what genes such as *Pax6* are doing by thinking that just because they are expressed in certain regions it is telling us something important? How can we use this approach as a strategy?

Rubenstein: There isn't a simple answer. However, some *Pax* genes are responsive to sonic hedgehog (Shh) as well as to bone morphogenetic proteins (BMPS), which tells us something about the position of some of these transcription factors with regard to patterning centres.

Purves: I'd like to bring the discussion back to the cortex, i.e. whether the cortex has antecedents or whether it has evolved in some other way. My view of evolution is that it always proceeds by tinkering, so my question is, what is the alternative to this tinkering?

Karten: Thirty years ago we viewed the mammalian neocortex as a totally novel structure—this was the underlying notion of 'neocortex'—and that what existed in non-mammals was a sort of laminated configuration, such as in the olfactory system. The specific sets of input and output connections involved in information processing characteristically defined in the studies of the mammalian cortex within the last 100 years were viewed as properties unique to mammals. It was argued until about 30 years ago that what we call cortex, in terms of its structure, constituents wiring and performance, was a novel evolutionary appearance. This is in striking contrast to what we would say about virtually any other part of the nervous system, or indeed any other part of the organism. What has now emerged is the realization that the neuronal constituents which make up the cortex have ancient histories. We can identify auditory and somatosensory neurons in birds or lizards for example, but they are in a different location and they don't look the way cortex looks. Therefore, are they truly new? We need to address this by first finding out whether there are any corresponding structures of a similar nature, and then seeing if the developmental transformations we are referring to can account for the evolutionary change. If this is the case, we would then say that the same constituents have just been shuffled around. If they have been tinkered with in this way, then we would want to know how. This is the challenge that some of us have dealt with in trying to understand the origins of cortex, i.e. neocortex is not new but has been around in one form or another as cells.

Puelles: There are several different layers of meaning at which we can interpret the word 'new', i.e. there may be new layers, changes in cell types or new fields. For instance, do lampreys have neocortical fields? We can discuss whether primordia such as the neural tube are new or similar to elements found in *Drosophila*. In this sense, we are dealing with evolutionary emergent phenomena. In theory, the same genetic bases can be duplicated and combined in different ways, and significant structural and functional novelty may arise in the course of time, but the basic question is whether there are any new genetic elements in morphogenesis.

Karten: This is an important point. There are two major levels at which we can address problems of homology, i.e. field homology and homology at the cellular level. I would like to ask Ann Butler to help us define those terms.

Butler: Field homology refers to the set structures derived from the same developmental field. For example, digits are homologous to each other as a set.

Karten: Is a neuromere a field? Or does it represent a group of identified neurons? That is, are they specified neurons or specified cells?

Butler: Yes, I would say that a neuromere is a field. It is a particular identifiable region of an embryo at a certain point in time. It would contain multiple sets of identified neurons.

Karten: That region can be identified in different clades, so would you then argue that they are homologous?

Butler: I would argue that the structures produced by two similar neuromeres are homologous to each other as a field homology, i.e. as a set of structures. Glenn Northcutt, however, has disputed this. He argues that if development proceeds further in one animal than in another, there are different levels of development, and this therefore invalidates the concept of field homology (Northcutt 1999).

Karten: I know how to recognize catecholaminergic amacrine cells of the retina. I look for the production of tyrosine hydroxylase in a particular zone within the retina at a certain stage of development. If we could deal with the problem of the evolution of brains at the level of the single cell, i.e. the identifiable neuron, then maybe this would be fairly easy to solve.

Puelles: You cannot look at cell type without also considering the position. You cannot say 'this cell is catecholaminergic and therefore it is an amacrine cell'. It depends where it is in the brain. If it is in the retina, it may be an amacrine cell; but if it is in the solitary nucleus then it may be something else.

Karten: You and John Rubenstein have recently been arguing for a revival in the concept of field homology, so please give us your definition.

Puelles: There are several theoretical definitions of the term 'field' in development. It is predominantly reserved at early stages for a set of homogeneous or non-homogeneous cells that are able to communicate with each other and have common boundaries that separate them from other surrounding sets of cells, with which they communicate less efficiently. This defines a causal subsystem within a larger system, where the prospective character states (cell fates) may find various equilibrium states within the same field along time and space parameters, but the whole is still causally interactive and largely causally independent from adjacent fields. The internal causal interaction secures the structural relative homogeneity of processes occurring within the field, but is also a motor for differentiation and subsequent variation. These fields usually arise by independization (boundary formation) within earlier more comprehensive fields, often preceded by an increase in cell population, though this is not strictly necessary. At later stages, the term 'field' is also used less strictly for the whole tissue domain thought to derive from one of the early histogenetic fields, independently of its final degree of regionalization and differentiation. This concept is less strict because secondary causal interactions between adjacent or distant early fields often need to be assimilated (i.e. afferent and efferent axonal projections and resulting trophic effects, or tangential neuronal migrations). The idea is that somehow the different field derivatives may undergo differential morphogenesis and evolution, but they still retain a common fundamental identity, because at a given early stage they shared similar precursors and thus they are derived from the same sets of cells (position within the overall *Bauplan*), which originally shared a given molecular constitution.

In evolution, the same field may give rise to many different field homologues, depending on the developmental interactive complexities superimposed upon the initial comparable field. I would like to propose that field homology is not only possible, but actually is the only sort of homology that can be postulated, once we have enough knowledge on the comparable parts. The concepts of 'isocortex', 'identifiable cell type' or 'potassium channel' imply also field homologies at different orders of magnitude, since we concentrate on the similar causal background and consequent structural similarity, momentarily disregard secondary diversification, and are equally dependent on positional context. Note how function remains an epiphenomenon due to independent variation of the context and is subject to either subtle, epistatic, or sudden catastrophic changes.

Wolpert: So, in wild-type *Drosophila*, would you say that the leg is homologous to the antenna?

Puelles: Not necessarily, because they occupy different initial positions in the *Bauplan*, which apparently confers a differential identity, independently of similarities in internal signalling. At a different level of analysis, they may be indeed comparable as serial appendages with a comparable morphogenetic programme for proximo-distal differentiation, which implies shared sets of genes. This seems to place the greatest weight of homology on position relative to the earliest developmental field (understood as precursor causal system); this may explain why heads are always heads and tails cannot be other than tails.

Wolpert: But the same communication pathways operate between the cells. The field is identical in the leg and in the antenna.

Butler: Ghiselin (1966) pointed out a number of years ago that it is important to specify (stipulate) homology. The antenna is homologous to the leg in an iterative sense, but it is not homologous as a developmental field. An example of a developmental field homology is the anterior thalamus, in which there is a single nucleus in fish and amphibians and multiple nuclei in amniotes. As a field homology those multiple nuclei in amniotes are homologous to the single anterior nucleus in fish and amphibians.

Reiner: I'm not the greatest fan of field homology, but I do have an example of where it can be used appropriately. Birds have 14 cervical vertebrae and mammals have seven. Which vertebra in birds is homologous to which particular one in the mammal? It is not possible to assign the various vertebrae; you have to revert to field homology and say these cervical vertebrae in birds are homologous as a field to the cervical vertebrae in mammals.

Wolpert: So, does the concept of homology in this situation help you?

Pettigrew: Emil Zuckerkandl made the point in 1994 (at the Society for Molecular Evolution conference in Costa Rica) that it is possible to argue, on the basis of the homeobox studies, that the bat wing and the insect wing are homologous. They share the same set of genes. My problem with the concept of homology is that people talk about it before they know the phylogeny, and therefore they inevitably go around in circles.

I would like to talk about the issue of timing. In order to choreograph a developmental pathway, you need to consider time as well as position. I wondered why Lewis Wolpert didn't refer to the fact that homeobox genes may represent clocks. McGrew et al (1998) have been working on this for the chicken *hairy* gene, which seems to be a transcriptional clock that doesn't involve proteins. This leads to another concept. We are all focusing on downstream targets, i.e. proteins and cells, but perhaps we should be thinking about the possibility that some of these developmental programmes operate at the genomic level. Perhaps John Mattick's idea of an RNA-type programme in the nucleus is relevant. I would also like to draw attention to Dennis Bray's work showing that if there is a network of proteins involving different pathways, there is a tremendous precision in time (Bray 1998). When more than a third of a signalling pathway is knocked out, a bacterium still has a chemotaxis time constant of exactly 1.5 seconds. There are many other examples where timing is absolutely crucial to development.

Wolpert: On the whole, developmental biologists don't spend much time on time. There is evidence in the nematode that certain cells are measuring time, but in general the timing of events reflects a cascade of gene activity. We have a model for the development of the chick limb in which the cells do measure time, and the reason why your digits are different from your humerus is because they have been in a particular region, the progress zone at the tip of the limb, for longer.

Papalopulu: Developmental biologists are aware of timing when they are looking at the concept of competence. When cells are exposed to an inducer, the cells respond only when they are pre-programmed to respond. The problem is that timing is a difficult issue to tackle.

Wolpert: But are those cells really measuring time?

Papalopulu: In general, cells respond within a narrow window. Beyond that window they may still respond, but they may give a different response. We don't really know what these cells are measuring, but it could be related to timing. The issues of timing and growth control are the two main issues that make structures different from each other.

References

Bray D 1998 Signaling complexes: biophysical constraints on intracellular communication. Annu Rev Biophys Biomol Struct 27:59–75

Ghiselin MT 1966 An application of the theory of definitions to systematic principles. Syst Zool 15:127–130

Levitan D, Greenwald I 1995 Facilitation of *lin12*-mediated signalling by *sel12*, the *Caenorhabditis elegans S182* Alzheimer's disease gene. Nature 377:351–354

McGrew MJ, Dale JK, Fraboulet S, Pouquié O 1998 The *lunatic fringe* gene is a target of the molecular clock linked to somite segmentation in avian embryos. Curr Biol 8:979–982

Northcutt RG 1999 Field homology: a meaningless concept. Eur J Morphol 37:95–99

Thoughts on the cerebellum as a model for cerebral cortical development and evolution

Karl Herrup

Department of Neuroscience and University Alzheimer Center of Cleveland, Case Western Reserve University, 10900 Euclid Avenue, Cleveland, OH 44120, USA

Abstract. This chapter explores the prospect of using the cerebellar cortex as a model for the development and evolution of the cerebral neocortex. At first, this would seem a nearly fruitless task given the readily apparent structural and functional differences between the two cortices. Cerebellum and cerebrum perform different associative tasks, the cellular 'circuit diagram' of the two structures is different, even the developmental sequences that give rise to the two structures differ markedly. Yet there are similarities between the structures at the conceptual level that are difficult to ignore. Both structures have a relatively simple modular circuitry and achieve their complexity by an increase in either the size or number of the modules. Both have massive commisures connecting the left and right halves of the structure. For the cortex this commisure is the obvious corpus callosum; the cerebellar commisure is made up of parallel fibres of the granule cells that pass freely across the midline. As they are thin and unmyelinated, the number of these crossing fibres may well exceed the number of the callosal axons by a significant amount. By far the most obvious similarity between cortex and cerebellum, however, is that they are both topologically sheet-like in structure. They are broad and wide in the two-dimensional plane of the pial membrane with a relatively modest thickness in the radial dimension. The question for this chapter then is whether these similarities, in particular the sheet-like organization are coincidental or indicative of larger themes that play deeper roles in the development and function of these two seemingly disparate brain regions.

2000 Evolutionary developmental biology of the cerebral cortex. Wiley, Chichester (Novartis Foundation Symposium 228) p 15–29

Development of the cerebellum

The cerebellar field is first defined in the early embryo shortly after the closure of the neural tube begins. In the mouse, this occurs at approximately embryonic day (E) 8 (shown diagrammatically in Fig. 1; for reviews see Wassef & Joyner 1997, Beddington & Robertson 1998, Martinez et al 1999). A transverse band of *Pax2* gene expression appears at the border of the mesencephalon and metencephalon. This is followed by similarly localized bands of *Fgf8* and *Wnt1* expression. A more

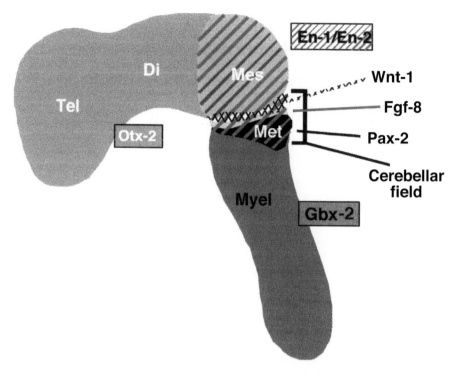

FIG. 1.   A diagrammatic representation of the transcription factors that specify the cerebellar field. The five major brain vesicles are indicated by the abbreviations over the embryo itself. The domains of transcription factors Otx-2 (grey), Engrailed (striped), Wnt-1 (crosshatched), Fgf-8 (light grey), Pax-2 (black) and Gbx-2 (dark grey) are indicated in their approximate positions relative to one another. Additional details can be found in the text. Di, diencephalic vesicle; Mes, mesencephalic vesicle; Met, metencephalic vesicle; Myel, myelencephalon or the rhombencephalic vesicle; Tel, telencephalic vesicle.

complex expression pattern of the Engrailed genes (*En1* and *En2*) follows with a peak of expression at the *Wnt1* band, a sharp decline on the posterior side and a more gradual decreasing gradient of gene expression on the anterior, mesencephalic side of the field. Recent experiments have identified additional players in this early scheme. The posterior extent of *Otx2* gene expression defines the anterior border of the cerebellar field while the anterior border of the hindbrain *Gbx2* expression appears to define the posterior border of the cerebellum.

After the pontine flexure forms, the cerebellar anlage is located in the roof of the fourth ventricle (Fig. 2). This position marks the cerebellum as an alar plate derivative, and suggests its categorization as a primarily sensory structure. The two principal neuronal cell types, the large neurons of the deep cerebellar nuclei (DCN) and the Purkinje cells of the cerebellar cortex, are the first to emigrate from

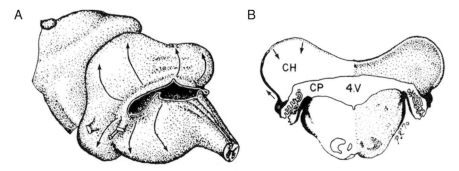

FIG. 2.   Migration pattern of the early cerebellar granule cells. (A) Three-dimensional view of the migratory paths of the granule cells over the surface of the embryonic cerebellum. The upward pointing arrows indicate the direction of migration of these precursors from the rhombic lip as they populate the external granule cell layer. (B) Transverse section through the embryonic cerebellar anlage showing the tangential surface migration (upward arrow on the left) followed by the centripetal migration from the external to the internal granule cell layer (two downward pointing arrows). The stippling on the right indicates the relative cell density of the large cerebellar neurons (Purkinje cells and neurons of the deep cerebellar nuclei). The drawing, by Pasko Rakic, was included in a review on neuronal migration (Sidman & Rakic 1973) and is reproduced here with permission.

the ventricular zone. Their migratory path is primarily radial, although the routes taken by some of the cells can appear anatomically torturous. The DCN neurons remain as nuclear groups in the cerebellar parenchyma; the Purkinje cells migrate further to populate the cerebellar plate. While this process is occurring, an unorthodox cellular migration takes place. Beginning in the lateral and posterior borders of the anlage, a group of *Math1*-positive cells (Ben-Arie et al 1997) leaves the rhombic lip and moves in an anterior and medial wave over the developing cerebellar surface, forming a superficial layer known as the external granule cell layer (EGL). These are the precursors of the granule cells of the internal granule cell layer (IGL). They multiply as they migrate and increase their numbers rapidly. Included in the tangentially migrating EGL population are a number of cells that secrete the large external protein, reelin (D'Arcangelo et al 1995). The exact function of reelin is unknown, but if it is disrupted by mutation the result is a massive failure of the early radial migration of most of the Purkinje cells (see below). This suggests that although the EGL cells might appear at first to be an unorthodox 'invasion' of cerebellar territory, they may in fact serve as an important organizing influence on the entire lamination process. This instructive role is emphasized further by the *unc5h3* mutation (Przyborski et al 1998). In homozygous *unc5h3*$^{-/-}$ mice, the cells of the EGL fail to detect the rostral border of the cerebellar anlage with the result that many granule cells invade the posterior inferior colliculus. This EGL ectopia is soon joined by an entire phalanx of Purkinje cells that layer just beneath the surface of the inferior colliculus. It is worth noting that a group of reelin-positive cells, the Cajal-Retzius cells, is also found in the early cortical plate of the cerebral cortex. Thus, the role of the invader as an organizing influence on cellular society is a theme that may be worth contemplating in studying the development of the cerebrum as well.

Granule cell migration normally ends with a cessation of cell division and a final, glial-guided centripetal migration through the developing molecular layer, past the Purkinje cell layer into the IGL. This migration is met by a smaller centrifugal migration from the white matter of a cell population consisting of DCN interneurons and Golgi II neurons, as well as stellate and basket cells of the molecular layer. We have recently shown (Maricich & Herrup 1999) that this final seemingly heterogeneous collection of neuron types arises from a single group of cells that appears in the waning ventricular zone (E13.5 in the mouse). The cells are marked by their expression of Pax2, now serving an apparently distinct function from its earlier 'cartographic' role. The Pax2-specified cell types share several common phenotypes: they use γ-aminobutyric acid (GABA) as a neuro-transmitter, and they are all short axon, local circuit interneurons. This Pax2/GABAergic interneuron correlation is also found in more caudal structures including the dorsal spinal cord. We have suggested that there is a shift in Pax2 function from one of specifying anatomical region to one of specifying neuronal cell phenotype (a regionalization, but not in a three-dimensional sense).

The origin of the stellate and basket interneurons has been debated over the years, Initial studies suggested that they arose from the EGL, but work with chick/quail chimeras was inconsistent with this view and their origin was proposed to be the ventricular zone (Hallonet et al 1990). Later retroviral studies extended this view by suggesting that a precursor population must exist in the postnatal cerebellum that gave rise to these molecular layer interneurons (Zhang & Goldman 1996). We have investigated this issue and demonstrated, both by BrdU incorporation as well as by the Pax2 immunolabelling of mitotic figures in the cerebellar white matter, that the Pax2-positive stellate and basket precursors are indeed dividing in the white matter as they migrate.

The genes required to build a cortex

Most of the known genes whose function is needed to specify the morphogenesis of cerebellum differ markedly from those required for cerebral cortical development. *Pax2*, *Fgf8*, *Wnt1*, *En1* and *En2* have no known effects on cerebral cortical development. Similarly, the pattern formation genes that lay out the field of the telencephalon are distinct from those in cerebellum (Rubenstein & Beachy 1998). The differences in cell type, cytoarchitecture and internal circuitry are undoubtedly a reflection of this lack of genetic overlap. Yet in the area of

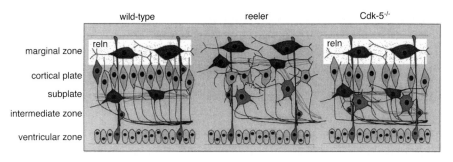

FIG. 3.   A comparison of early cortical plate development in wild-type, *reeler* and *Cdk5*-deficient mice. In wild-type animals the preplate cells (dark grey) are split into two layers, an upper marginal zone containing Cajal-Retzius cells that produce reelin and a lower subplate. The split occurs concurrent with the arrival of the early born cells of layer VI (light grey) followed by the cells of layer V (medium grey) and so forth. In *reeler* mice, reelin is absent and the migrating cortical plate cells cannot split preplate. The cells remain instead as a single superficial layer known as the superplate. In *Cdk5* mutants, reelin is present and the earliest cells (layer VI) do split the preplate. Subsequent migrants are seemingly unable to pass the subplate however and stack up beneath it in a layer termed the underplate (Gilmore et al 1998). reln, reelin.

neuronal migration, it would appear that there is a significant overlap in the mechanisms used by the two brain regions to produce a laminated cortical structure. Human cortical cell migration syndromes such as Zellweger's (Evard et al 1978) and lissencephaly of the Miller–Dieker type (Hirotsune et al 1998) also affect the developmental migration of cerebellar neurons. Similarly, many mouse mutations that were isolated because of their ataxia (and related cerebellar abnormalities) have cell migration defects that affect both cerebral as well as cerebellar cortices in similar ways.

The best known mouse mutations of this type are *reeler*, *yotari* and *scrambler*. *reeler* is caused by a mutation in the gene encoding reelin, a large external protein secreted by the Cajal-Retzius cells of the early preplate. In the cerebral cortex of reelin-deficient animals, the first-born (layer VI) cells migrate to, but do not split, the preplate as normally occurs in wild-type animals (Fig. 3). Instead this early cortical lamina remains as a superplate beneath which all subsequent waves of migrating cortical neurons collect. In reelin-deficient cerebellum, most Purkinje cells remain lodged deep in the cerebellar parenchyma, surrounding the relatively normally positioned neurons of the DCN. The cells of the EGL appear to complete a normal tangential migration, and the final centripetal migration to the IGL has no reported defects, but is significantly reduced in size due to the overall cytoarchitectonic abnormalities. Curiously, unlike cerebral cortex, a small but significant number of *reeler* Purkinje cells successfully complete migration to the Purkinje cell layer (Mariani et al 1977). The *scrambler* and *yotari* mutations, caused by a defect in the mouse *disabled* gene (*Mdab1*), demonstrate an identical phenotype in both cerebral and cerebellar cortex (Sheldon et al 1997).

A second group of mutations that disrupt migration in both cerebral and cerebellar cortex are those that interfere with the activity of the protein kinase known as cyclin-dependent kinase-5 (encoded by *Cdk5*; Ohshima et al 1996) or its activator protein, p35 (Chae et al 1997, Kwon & Tsai 1998). The phenotype of the cerebral cortex of the $cdk5^{-/-}$ mouse is nearly identical to that of *reeler* with one important distinction: the cells of the deeper cortical layers (layer VI) successfully split the preplate. The cells of the later, more superficial cortical layers stack up below the subplate in a configuration similar to *reeler* and *scrambler* mice. While this cerebral phenotype might be viewed as milder than that of *reeler* and *scrambler*, the situation in the cerebellum is reversed. None of the $cdk5^{-/-}$ Purkinje cells successfully migrate to the cerebellar cortex (Ohshima et al 1999). This suggests that the $cdk5^{-/-}$ mutation causes a more severe arrest of Purkinje cell migration than either the *reeler* or *scrambler* mutation. Further, while no granule cell migration defect is reported in *reeler/scrambler* mice, *Cdk5*-deficient mice have significant cell autonomous defects in granule cell migration. The *p35* defects, while similar in kind to those of the $cdk5^{-/-}$ mice, appear to represent a subset of the latter (Chae et al 1997, Kwon & Tsai 1998).

These examples, two in human, five in mouse, suggest that both cerebral and cerebellar cortex rely on common molecular mechanisms of cell migration to construct the sheet-like topology of their adult structure. This observation is all the more intriguing because there are many other mutations that affect the development and function of these two structures uniquely. The common mutations all appear to affect one or another aspect of cell migration. It is of course possible that this is merely coincidence, but it should alert us to the possibility that these migration patterns are a common feature of cortical development.

The evolution of the cerebellum

In a chapter of this length, it is not feasible to do justice to the complex topic of cerebellar evolution. Rather, a few observations are presented here for consideration in the context of the relevance of the evolution of the cerebellar cortex to that of the cerebral cortex.

Perhaps the most basic observation of all is that the cerebellum as a laminated cortex evolved in vertebrates well before its more anterior cousin, the cerebrum. Thus birds, reptiles, amphibia and several species of fish have significant layered cerebella at the midbrain/hindbrain junction, with most of the major cell types present. This earlier evolution may relate to the relative simplicity of the layering in the cerebellum compared to the cerebrum, although this is pure speculation. The cerebellum is generally believed to have evolved from cells subserving the lateral line organ of fish, a somatosensory array that alerts the organism to the movement of the surrounding water.

The comparative anatomy of the expression of the glycolytic enzyme, aldolase C, has been used to suggest a set of steps in the evolution of the cerebellum (Lannoo et al 1991). Aldolase C is a glycolytic enzyme that catalyses the cleavage of fructose-1,6-bisphosphate and is the antigen recognized by the monoclonal antibody, Zebrin II (Ahn et al 1994). Its location in the brain was first described in the rat cerebellum where it uniquely labels a subset of cerebellar Purkinje cells (Hawkes et al 1993). The pattern of the Zebrin-positive cells defines a reproducible pattern of seven sagittal bands intercalated by seven unstained interbands. The bands run nearly continuously from anterior to posterior cerebellum and a series of tracing studies has shown that the boundaries defined by this staining pattern are nearly congruent with the anatomical projection pattern of the climbing fibres. A less perfect registration of the cerebellar mossy fibre afferents is also described. The 7+7 band pattern has been reported in mammalian cerebella of all sizes ranging from mouse to human. Thus, unlike cerebral cortex, in which expansion involves the addition of new cytoarchitectonic areas, the cerebellar pattern appears invariant. In birds, although the hemispheres of the cerebellum are poorly developed, the vermal pattern of Zebrin II bands is retained. By contrast, in teleost fish, cerebellar function is parcelled out into different regions. Each region contains Purkinje-like cells, but the Zebrin antibody now stains in an all-or-none fashion. For example, all of the Purkinje cells in the corpus cerebella are Zebrin-positive while all of the cells in the lateral valvula are Zebrin negative. The suggestion is that the cerebellum of birds and mammals evolved by the interdigitation of these initially separate cell groups. That the interdigitation was part of the process that led to the emergence of a cortical cytoarchitecture to the cerebellum is a topic that would appear ripe for pursuit.

Conclusions and food for thought

The cerebellum and cerebrum have arisen from different primitive brain regions at different times in evolution and apparently for different reasons. It seems plain on both developmental and evolutionary grounds that the characteristic sheet-like cortical architecture of the two regions evolved separately in the two areas. The speculation would be that the increase in the processing power of the architectural arrangement of cells was advantageous in both cases, but the exact cellular solutions to achieve this end appear to have been quite different.

Our Pax2 study has validated the suggestion of Zhang & Goldman (1996) that the GABAergic interneurons of the molecular layer originate from migratory mitotic precursors located in the folial white matter (Maricich &

Herrup 1999). This post-ventricular neurogenic region suggests an analogy with the subventricular zone of the cerebral cortical neuroepithelium. In both cases, a late-arriving population of interneurons is born in a site removed from the traditional ventricular zone. In the cerebrum, the site remains deep within the parenchyma of the telencephalic vesicle while in the cerebellum, the stellate/basket precursors migrate some distance into the cortex before they cease division. The reasons for and advantages of establishing such a secondary germinative zone are unclear, but the analogy suggests that comparative studies between the two cell groups in cerebrum and cerebellum may well be worth pursuing.

A review of cerebellar development and evolution emphasizes the role of non-radial cell movements as key events in both processes. The interdigitation of the Zebrin-positive and Zebrin-negative cell groups during the evolution of cerebellum from fish to birds is a potential example of this. A more definitive example would be the massive tangential migration of the external granule cells during cerebellar development. The *reeler*, *scrambler* and *unc5h3* mutations suggest that this invasion of *Math1*-positive cells from the rhombic lip is not merely an exercise in space filling. The migration and final positioning of the large Purkinje cell neurons would appear to be highly dependent on these invaders. Given the evidence to date, it seems likely that this effect is mediated through the reelin pathway, but this conclusion is far from proven. A worthwhile experiment in this regard would be to create *reeler/unc5h3* double mutants. The predicted outcome would be that the EGL cells would still overshoot the cerebellum and enter the colliculus, but the Purkinje cells should fail to follow.

The mix of radial and non-radial migrations that populate the cerebral cortex is far more heavily skewed toward the radial. Tangential migrations have been documented in cerebrum, however, and the message from the cerebellum is that these migrations should be examined carefully for instructive cues that guide rather than simply participate in cortical lamination. The Cajal-Retzius cells would be an obvious candidate for this function given their secretion of reelin and their clear role in cortical migration. Yet, unlike the reelin-secreting cells of the EGL, these primitive neurons do not appear to reach cortex tangentially, but rather directly from the early telencephalic neuroepithelium. Another population that might be considered as a source of lamination cues are the cells that migrate into neocortex from the ganglionic eminence. These cells are lost in *Dlx*-deficient mice suggesting that this transcription factor may act to retain their identity during migration much as the EGL cells appear to use *Math1* expression.

In the final analysis, there are many deep differences between cerebellum and cerebrum in both their development and evolution. Yet the similarities of structure and apparent reliance on common migration tools to achieve the laminated appearance suggest that the 'little brain' might none the less have useful hints to guide the study of cerebral cortex.

Acknowledgements

This work was supported by grants from the National Institutes of Health (NS20591, NS18381, and AG08120) and the Blanchette Hooker Rockefeller Foundation.

References

Ahn AH, Dziennis S, Hawkes R, Herrup K 1994 The cloning of *zebrin II* reveals its identity with aldolase C. Development 120:2081–2090

Beddington R, Robertson E 1998 Anterior patterning in mouse. Trends Genet 14:277–284

Ben-Arie N, Bellen HJ, Armstrong DL et al 1997 *Math1* is essential for genesis of cerebellar granule neurons. Nature 390:169–172

Chae T, Kwon YT, Bronson R, Dikkes P, Li E, Tsai LH 1997 Mice lacking p35, a neuronal specific activator of Cdk5, display cortical lamination defects, seizures, and adult lethality. Neuron 18:29–42

D'Arcangelo G, Miao GG, Chen SC, Soares HD, Morgan JI, Curran T 1995 A protein related to extracellular matrix proteins deleted in the mouse mutant *reeler*. Nature 374:719–723

Evard P, Caviness V, Prats-Vinas J, Lyon G 1978 The mechanism of arrest of neuronal migration in the Zellweger malformation: an hypothesis based upon cytoarchitectonic analysis. Acta Neuropath 41:109–117

Gilmore EC, Ohshima T, Goffinet AM, Kulkarni AB, Herrup K 1998 Cyclin-dependent kinase 5 deficient mice demonstrate novel developmental arrest in cerebral cortex. J Neurosci 18:6370–6377

Hallonet MER, Teillet MA, Le Douarin NM 1990 A new approach to the development of the cerebellum provided by the quail-chick marker system. Development 108:19–31

Hawkes R, Blyth S, Chockkan V, Tano D, Ji Z, Mascher C 1993 Structural and molecular compartmentation in the cerebellum. Can J Neurol Sci (suppl 3) 20:S29–35

Hirotsune S, Fleck MW, Gambello MJ et al 1998 Graded reduction of Pafah1b1 (Lis1) activity results in neuronal migration defects and early embryonic lethality. Nat Genet 19:333–339

Kwon YT, Tsai LH 1998 A novel disruption of cortical development in *p35(−/−)* mice distinct from *reeler*. J Comp Neurol 395:510–522

Lannoo MJ, Ross L, Maler L, Hawkes R 1991 Development of the cerebellum and its extracerebellar Purkinje cell projection in teleost fishes as determined by zebrin II immunocytochemistry. Prog Neurobiol 37:329–363

Mariani J, Crepel F, Mikoshiba K, Changeuz JP, Sotelo C 1977 Anatomical, physiological and biochemical studies of the cerebellum from *reeler* mutant mouse. Phil Trans R Soc Lond B Biol Sci 281:1–28

Maricich S, Herrup K 1999 *Pax-2* expression defines a subset of GABAergic interneurons and their precursors in the developing murine cerebellum. J Neurobiol 41:281–294

Martinez S, Crossley P, Cobos I, Rubenstein JLR, Martin GR 1999 FGF8 induces formation of an ectopic isthmic organizer and isthmocerbellar development via a repressive effect on *Otx2* expression. Development 126:1189–1200

Ohshima T, Ward JM, Huh CG et al 1996 Targeted disruption of the cyclin-dependent kinase 5 gene results in abnormal corticogenesis, neuronal pathology and perinatal death. Proc Natl Acad Sci USA 93:11173–11178

Ohshima T, Gilmore E, Longenecker G et al 1999 Migration defects of *cdk5−/−* neurons in the developing cerebellum is cell autonomous. J Neurobiol 19:6017–6026

Przyborski SA, Knowles BB, Ackerman SL 1998 Embryonic phenotype of *Unc5h3* mutant mice suggests chemorepulsion during the formation of the rostral cerebellar boundary. Development 125:41–50

Rubenstein JLR, Beachy PA 1998 Patterning of the embryonic forebrain. Curr Opin Neurobiol 8:18–26

Sheldon M, Rice DS, D'Arcangelo G et al 1997 *scrambler* and *yotari* disrupt the *disabled* gene and produce a *reeler*-like phenotype in mice. Nature 389:730–733

Sidman RL, Rakic P 1973 Neuronal migration, with special reference to developing human brain: a review. Brain Res 62:1–35

Wassef M, Joyner A 1997 Early mesencephalon/metencephalon patterning and development of the cerebellum. Perspect Dev Neurobiol 5:3–16

Zhang L, Goldman JE 1996 Generation of cerebellar interneurons from dividing progenitors in white matter. Neuron 16:47–54

DISCUSSION

Parnavelas: But the subventricular zone in the developing cerebral cortex does not contain precursors that contribute to neuronal population of the cortex.

Karten: Could you specify what you mean by subventricular zone? Because we may be using the concept of subventricular zone differently.

Parnavelas: The subventricular zone is distinguished as a separate layer of cells overlying the germinal ventricular zone. It first appears in the developing cortex as the ventricular zone begins to diminish in prominence. In rodents, the subventricular zone expands greatly during late gestation, and in early postnatal life it comes to reside adjacent to the lateral ventricle and just underneath the formative white matter (Sturrock & Smart 1980). In the postnatal brain, this zone may be seen as a mosaic of glia progenitors that give rise to cortical astrocytes and oligodendrocytes, of multipotential progenitors, of neuronal progenitors that produce a population of olfactory bulb neurons, and of a pool of stem cells. However, it does not contain progenitors of cortical neurons.

Herrup: Glial cells are produced from the dividing cells in the ventricular zone, although the Pax2-positive subset only give rise to neurons. As you point out, neurons migrate from the subventricular zone to the olfactory bulb, but the mix of the two cell types is different in the two structures. I was struck by this extraventricular site of cell genesis, and I wondered whether there might be homologies.

Karten: Not everyone agrees with John Parnavelas' definition of the subventricular zone.

Parnavelas: It's not my definition, it's the one given by the Boulder Committee (1970).

Rakic: In 1970 in was difficult to ascertain the nature of the subventricular zone, and whether it produces neurons as well as glia. In 1975 we suggested that in primates it also generates neurons, and in particular the stellate cells destined for the more superficial cortical layers (Rakic 1975). However, the majority of researchers in the field agree with John Parnavelas, i.e. that it produces only glial cells. The border between these proliferative zones is also difficult to define. I like the operational definition. Dividing cells in the ventricular zone are those that are attached to the ventricular surface, whereas cells in the subventricular zone divide *in situ*. They are not attached to the ventricular surface and are therefore more prone to lateral movement. Therefore, this definition is based on cell behaviour, which one can study in slice preparations. In contrast, the definition in 1970 was based on morphology.

Karten: Unfortunately, we don't have a presentation on tangential migration, but John Rubenstein recently addressed this issue, so I would like to ask him to comment on this.

Rubenstein: Our point of view is that during prenatal development the subventricular zone may well be a site of neurogenesis for neurons that migrate to the cerebral cortex. There are at least two types of tangenially migrating interneurons that migrate from the basal ganglia to cortical areas. The first migrates within the marginal zone, and the other appears to migrate in the intermediate zone. I can't be sure whether or not some of these latter cells are in the subventricular zone, especially late in gestation.

Rakic: The only evidence for this may be the enlarged portion of the subventricular zone. It has yet to be proved that neurons originating in this zone will become projection neurons.

Parnavelas: In my view, those neurons are in the intermediate zone and not in the subventricular zone. The source of confusion lies in the shape of the cells in the subventricular zone, i.e. they tend to be horizontally orientated in a similar fashion as the migratory cells in the intermediate zone.

Bonhoeffer: What do you know about the forces that drive tangential migration?

Herrup: I know very little. The *unc5h3* mutation suggests that the netrins and their receptors are involved in providing a stop signal to the tangential dimension of the migration of the external granule layer (EGL) cells. If you follow their path in the *unc5h3* mutant, you find that they enter the inferior colliculus. We don't know why they go there, nor why they don't migrate caudally into the brainstem.

Puelles: In my opinion, this interpretation is too simplistic, because in normal mice there is no direct connection between the cerebellum and the inferior colliculus, the whole isthmus being intercalated in between. This suggests that in this mutant there is a patterning defect that eliminates the isthmus altogether, in addition to a migratory defect.

Molnár: I would like to suggest that perhaps the cerebellum is a good model for separating its different parts with different evolutionary origins. Karl Herrup didn't discuss whether the old and novel extensions of the cerebellum have different mechanisms of development, but I would like to know what triggers these extensions, and what are the underlying differential gene expression patterns in the different parts of the cerebellum (i.e. archicerebellum, paleocerebellum, neocerebellum)?

Herrup: That's a good question. I would describe the main differences as differences of geography. Unlike the areas of cerebral cortex, when the different parts of the cerebellum are built, the microarchitecture remains the same.

Molnár: Is it possible that the mechanisms underlying the different extensions are duplicated, so that the new areas dealing with functions such as language, and perhaps cognition, have the same developmental programmes and building principles, yet they can perform novel tasks?

Herrup: The early developmental programmes differ, which is where the distinction between the anterior, posterior and floculonodular lobes come from. We don't know the nature of these differences. We know that mutations of the *meandertail* gene clearly define a genetic difference between the anterior lobes of the cerebellum and the posterior and floculonodular lobes. I know of no mutation that does the reverse.

Rakic: One difference between the cerebrum and the cerebellum is that whatever the cerebellum does in the cat or mouse, it also does in humans; whereas in the cerebrum, novel and functionally different areas are introduced during evolution. In addition, the ratio of granular cells to Purkinje cells varies among different species, which probably reflects differences and elaborations of function, whereas the situation in the neocortex is different because new areas assume different functions.

Reiner: I would like to follow up on the points raised about the evolutionary aspects of the cerebellum versus the cerebral cortex. Although parts of the telencephalon in birds and reptiles may not resemble cerebral cortex, they do perform the same kinds of function. The cerebellum has therefore independently expanded in birds and reptiles compared to mammals in accordance with the expansion of the parts of the telencephalon that are devoted to cerebral cortex-like functions.

Molnár: Recent literature (Desmond & Fiez 1998) does not support the suggestion that the function of the cerebellum is similar in rats, monkeys and humans. The primate cerebellum has an important function in language, and perhaps even in cognition, whereas this is not the case for rats.

Levitt: I'm not sure that we say this definitively. The 'so-called' new cognitive areas in primate cerebellum simply haven't been analysed in rodents or the chicken. For all we know there are cognitive components of every motor output, and there are representations in the cerebellum that reflect this.

Molnár: Are the potentials for cognition functions present in the rat?

Levitt: My guess is that a chicken plans a movement, and part of that planning process involves circuitry contained within specific domains of the cerebellum, as has been shown in humans and non-human primates. I just don't know whether these experiments have been done in chickens or rats.

Karten: In regard to the cerebellum, and perhaps also in the hippocampus, we are dealing with a defined and limited number of neurons. In many other systems, e.g. the retina, we can also define the number of neurons based on certain objective criteria pertaining to morphology, position and biochemical properties. Can we even begin to do this for the cortex? How many types of neurons are required to build a cerebral cortex?

Rakic: This probably depends on the definition or type of an individual neuron, i.e. whether neurons containing different combinations of peptides are classified as being of different types. If you believe this, then the numbers will be high. However, if you classify the neurons on the basis of morphology alone, then the numbers will be much lower.

Karten: But in the olfactory bulb, for example, there are defined numbers of cells, so we can talk about the evolutionary constancy of variability. For instance, if I look for a dopaminergic periglomerular cell in the olfactory bulb, I will find it in every vertebrate I look at. We are talking about how the layered structure of the cerebral cortex is built, but from where are the constituents of this layered structure derived?

Parnavelas: The work of the early neuroanatomists (Lorente de No 1949), and others since then, suggests that there are no clear-cut differences between the disposition and morphology of pyramidal cells in the rodent cortex and in the cortices of higher mammals. On the other hand, the overall dendritic and axonal morphology of the non-pyramidal cells in rodents appear simpler than in cat, monkey and human.

Karten: I would say that there are many classes of pyramidal cells because they have different biochemistries and different connections. This raises an important point, i.e. what criteria should we use to define individual cell types? If we say that there are many different classes of pyramidal cells, how can we say that they don't vary?

Pettigrew: I have asked many people whether they would expect to see any differences between the behaviours of layer III pyramidal cells in V1, V2, V3 and V5. Most people said that they wouldn't, which is the wrong answer. Elston & Rosa (1998) injected layer III pyramidal cells in these regions, and found that the number of spines and the complexity increases fourfold as you go 'up' the visual hierarchies represented by the dorsal and ventral streams. Most people assumed that layer III pyramidal cells conformed to a particular archetype, which has turned out not to be the case. Therefore, I would say that the problem Harvey Karten has raised is partly quantitative.

Karten: Do we have any agreement on whether evolution is operating at the level of different cell types? Or is there a more vague property that defines cell groups?

Puelles: Evolution is supposed to work on developmental processes (largely quantitative changes, less frequently qualitative change), through selection of ultimate functional capabilities, and not directly on different cell types. A cell type that we are tempted to define simplistically as 'the same' may in fact arise through alternative developmental pathways in different brain positions (see Puelles & Verney 1998).

Karten: The gene may be the vehicle for evolutionary change, but in a different sense selection operates on the adult animal, i.e. on the phenotype, and the phenotype is a manifest expression of the gene. The question is, how does it make this transition? And what is the nature of the continuous identity that the gene is concerned with preserving or modifying?

Herrup: The experiment I would like to do is to remove the layer III pyramidal cell from one of the higher areas of visual cortex and transplant into one of the primary visual areas, and ask whether it will take on the form of the lower areas, i.e. is it the cell's own genome that is causing the complexity or is it its environment that's inducing the complexity?

Molnár: I have another example of these sorts of differences. If you label cells in layer V in the adult rodent cortex by cell filling, you will find that some of them have apical dendrites that reach layer I and some don't. Until the end of the first postnatal week however, all layer V cells have similar morphology, all with apical dendrites ending in terminal tufts in layer I. Koester & O'Leary (1992) and Kasper et al (1994) showed that these two cell types have different projections: the ones that lose their apical tufts during development project to the contralateral hemisphere; and the ones that keep the apical tuft project to the spinal cord and the superior colliculus. These cell types cannot be found in the cat or monkey, in which stellate cells are responsible for this function. Therefore, even at the level of the individual cell, we have to be careful that we identify the projections and then put them into specific categories.

Karten: Another question is what constitutes the essential property of identity? Are there different types of Purkinje cells?

Herrup: Yes, there's at least two.

Karten: On what basis can they be differentiated?

Herrup: On their aldolase C content.

Pettigrew: On that point, you mentioned that inferior olive projects to the Zebrin-positive cells. Is the implication that the aldolase C-negative cells don't receive projections from the inferior olive?

Herrup: No. The projection to the inferior olive from any one location tends to be organized in sagittal bands. If you inject the entire olivary complex on both sides you will fill the cerebellar cortex. However, if you do a small injection to define those bands, you find that the borders of the bands tend to respect the borders defined by the Zebrin stain. Zebrin is a reflection of a larger modular architecture of the cerebellum. The same modules are revealed by cell death in a variety of mutant conditions, and it can be revealed by other agents such as cytochrome oxidase.

Karten: One current hot area is the field of stem cells. It is possible that in the future we will be able to regulate the sequence of operations that result in specification of a particular cell type. How far away are we in being able to answer what specifies Purkinje cell, for example, from a neural stem cell?

Herrup: I don't think we are even close.

References

Boulder Committee 1970 Embryonic vertebrate central nervous system: revised terminology. Anat Rec 166:257–261

Desmond JE, Fiez JA 1998 Neuroimaging studies of the cerebellum: language, learning and memory. Trends Cognit Sci 2:355–362

Elston GN, Rosa M 1998 Morphological variation of layer III pyramidal neurons in the occipitotemporal pathway of the macaque visual cortex. Cereb Cortex 8:278–294

Kasper EM, Lübke J, Larkman AU, Blakemore C 1994 Pyramidal neurons in layer 5 of the rat visual cortex. III. Differential maturation of axon targeting, dendritic morphology, and electrophysiological properties. J Comp Neurol 339:495–518

Koester SE, O'Leary DDM 1992 Functional classes of cortical projection neurons develop dendritic distinctions by class-specific sculpting of an early common pattern. J Neurosci 12:1382–1393

Lorente de No R 1949 Cerebral cortex: architecture, intracortical connections, motor projections. In: Fulton JF (ed) Physiology of the nervous system. Oxford University Press, Oxford, p 288–330

Puelles L, Verney C 1998 Early neuromeric distribution of tyrosine-hydroxylase-immunoreactive neurons in human embryos. J Comp Neurol 394:283–308

Rakic P 1975 Timing of major ontogenetic events in the visual cortex of the rhesus monkey. In: Buchwald NA, Brazier M (eds) Brain mechanisms in mental retardation. Academic Press, New York, p 3–40

Sturrock RR, Smart IH 1980 A morphological study of the mouse subependymal layer from embryonic life to old age. J Anat 130:391–415

Radial unit hypothesis of neocortical expansion

P. Rakic

Section of Neurobiology, Yale University School of Medicine, New Haven, CT 06510, USA

Abstract. The more than 1000-fold increase in the cortical surface without a comparable increase in its thickness during mammalian evolution can be explained in the context of the radial unit hypothesis of cortical development. Cortical expansion results from changes in the proliferation kinetics of founder cells in the ventricular zone that increase the number of radial columnar units without significantly changing the number of neurons within each unit. Thus, regulatory genes that control the timing (onset/rate/duration) and mode (symmetrical/asymmetrical) of cell divisions and the magnitude of programmed cell death (apoptosis) in the ventricular zone determine the number of cortical cells in a given species. The migration of postmitotic cells and their allocation into appropriate positions within the cortex is radially constrained by glial scaffolding and thereby creates an expanded cortical plate in the form of a sheet. Several families of genes and morphoregulatory molecules that control the production, migration and deployment of neurons within the developing cortical plate are being identified and their functions tested *in vitro* and in transgenic animals. The results provide a hint of how mutation of genes that regulate the early stages of corticogenesis may determine the species-specific size and basic organization of the cerebral cortex that sets the stage for the formation of the final pattern of its synaptic connections that can be validated through natural selection.

2000 Evolutionary developmental biology of the cerebral cortex. Wiley, Chichester (Novartis Foundation Symposium 228) p 30–45

The truism that nothing in biology makes sense unless viewed in light of evolution is perhaps most obvious in developmental neuroscience. It is generally agreed that the mechanisms underlying expansion of the cerebral cortex and its subdivision into functionally distinct areas is central to our understanding of the limits and potential of our cognitive capacity. As evident at this symposium, most studies of cortical evolution have been concerned with its origin and parcellation into functional areas. The genetic and cellular mechanisms by which the neocortex has expanded in the number of cells and surface area have not been equally well explored. Recent advances made in understanding the regulation of cell cycle and cell death open an opportunity to explore this challenging issue. I will outline here a model of the possible developmental mechanisms that may underlie the large expansion of cortical surface with relatively minor changes in its thickness.

A consistent feature of the adult cerebral cortex is the organization of its neurons into orderly horizontal and vertical arrays, which form anatomically and physiologically distinct laminar and columnar compartments. The columnar organization consists of an array of iterative neuronal groups (called interchangeably columns or modules) that extend radially (perpendicular to the pial surface) across cellular layers II to VI (e.g. Mountcastle 1997). The cells within a given column are stereotypically interconnected in the vertical dimension, share extrinsic connectivity and, hence, act as basic units subserving a set of common static and dynamic functions. In general, the larger the cortex in a given species, the larger the number of participating columnar units.

The increase in cerebral surface among living mammals can be illustrated by the ratio of the size of the neocortical surface between mouse, macaque monkey and human, which is approximately 1:100:1000, respectively (Blinkov & Glezer 1968). Since the thickness remains relatively steady, the initially smooth cortical surface becomes progressively more convoluted. Thus, an increase in size by expanding surface area is a primary prerequisite for cortical evolution. How may this have occurred at the cellular level? Which genes are involved? In order to provide a working model of cortical expansion, I will first describe early developmental events, such as the mode of cell production, pattern of neuronal migration, and emergence of laminar and modular organization in the cerebral cortex, including the timing of these events in mouse, macaque monkey and human. After that, I will propose a model of how genes that control the mode of cell proliferation and cell death at a critical stage of development can account for the changes in size of the cortex during evolution.

Critical cellular events

It is important to recognize that the species-specific size of the cortex is determined early, before neuronal connections have been established. In all, mammalian species of cortical neurons are generated in the ventricular and subventricular zones situated near the surface of the cerebral ventricle (see Fig. 1 and Rakic 1988, for review). Postmitotic cells produced in succession within these zones migrate across the intermediate and subplate zones before entering the developing cortical plate (Fig. 1 and Rakic 1972, 1974, 1981). The migrating cells are guided by glial scaffolding consisting of elongated, non-neuronal elements— radial glial cells—which express specific biochemical properties and exist only during the phase of neuronal migration (reviewed in Rakic 1997, Cameron & Rakic 1991). Although some classes of postmitotic cells originating from the subventricular zone and ganglionic eminence do not obey the constraints imposed by radial glial scaffolding (e.g. Rakic et al 1974, Luskin et al 1988, Tan et al 1998, Misson et al 1991, Ware et al 1999, Rakic 1990, Rubenstein et al 1999, Lois et al 1996) most neurons generated in the ventricular zone migrate radially to their destination (Rakic 1972, 1990). Recent re-evaluation of retroviral lineage studies indicates that distribution of clonally related cells in the rodent cortex are in harmony with the radial unit hypothesis (Ware et al 1999, Tan et al 1998, Rakic 1995a). Even in the large, convoluted primate cerebrum, clones of neurons in the

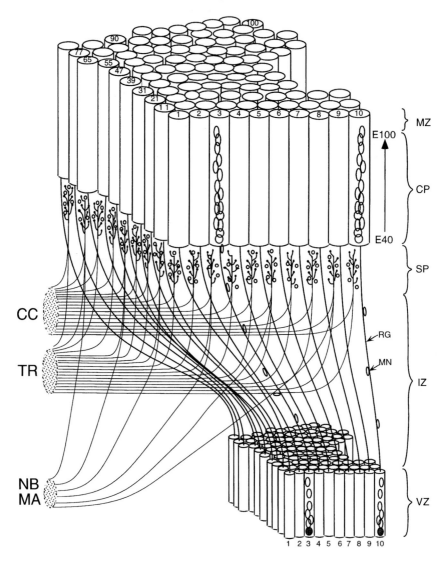

cortex remain in radial alignment (Kornack & Rakic 1995). The postmitotic neurons that are confined to the radial pathway exhibit affinity for elongated glial fibres and are termed 'gliophilic', in contrast to 'neurophilic' cells which move preferentially along a tangentially oriented axonal pathway (Rakic 1990). The surface molecules that provide differential adhesion between migrating neurons and glial fibres is being actively investigated (reviewed in Rakic et al 1994, Rakic 1997, Komuro & Rakic 1998, Pearlman et al 1998), but in the context of this chapter, it is sufficient to state that glial scaffolding is an essential prerequisite for building a cerebral cortex, as arrays of radial columns intersected by horizontal layers of isochronously generated neurons.

Radial unit hypothesis

The manner by which postmitotic cells migrate and become deployed in the three-dimensional matrix of the developing cortical plate is crucial for understanding how the neocortex expands in the form of a sheet rather than a lump as, for example, neostriatum. As reviewed above, the waves of postmitotic neurons generated within the same site in the ventricular zone arrive successively at the cortical plate, where they pass by each other and become arranged vertically in the form of cell stacks, named ontogenetic or radial columns (Rakic 1978, 1988). Thus, the radial unit consists of cells that originate from several clones that share their birthplace at the same spot in the ventricular zone, migrate along a common pathway and finally settle within the same column (Fig. 1). This organization

---

FIG. 1.   A three-dimensional illustration of the basic developmental events and types of cell–cell interactions occurring during the early stages of corticogenesis, before formation of the final pattern of cortical connections. This cartoon emphasizes radial migration, a predominant mode of neuronal movement, which, in primates, underlies the elaborate columnar organization of the neocortex. After their last division, cohorts of migrating neurons (MN) first traverse the intermediate zone (IZ) and then the subplate zone (SP), where they have an opportunity to interact with 'waiting' afferents arriving sequentially from the nucleus basalis and monoamine subcortical centres (NB, MA), from the thalamic radiation (TR) and from several ipsilateral and contralateral corticocortical bundles (CC). After newly generated neurons bypass the earlier generated ones situated in the deep cortical layers, they settle at the interface between the developing cortical plate (CP) and the marginal zone (MZ) and, eventually, form a radial stack of cells that share a common site of origin but are generated at different times. For example, neurons produced between embryonic day (E) 40 and E100 in radial unit 3 follow the same radial glial fascicle and form ontogenetic column 3. Although some cells, presumably neurophilic in nature of their surface affinities, may detach from the cohort and move laterally, guided by an axonal bundle (e.g. horizontally oriented, black cell leaving radial unit 3), most postmitotic cells are gliophilic, e.g. have affinity for the glial surface and strictly obey constraints imposed by transient radial glial scaffolding (RG). This cellular arrangement preserves relationships between the proliferative mosaic of the ventricular zone (VZ) and the corresponding proto-map within the SP and CP, even though the cortical surface in primates shifts considerably during a massive cerebral growth encountered in mid-gestation (for details see Rakic 1988).

enables translation of two-dimensional positional information contained within of the mosaic of the proliferative zone into three-dimensional cortical architecture: the X and Y axis of cell position within the horizontal plane is provided by the site of cell origin; whereas the Z axis along the depth of the cortex is provided by the time of its origin. Radial columns are particularly prominent in the primate cerebrum, easily recognized in histological sections cut across the cortical plate during mid-gestation in both monkey and human. Although the relation between ontogenetic and functional columns of the adult cortex (Mountcastle 1997) remains to be defined, the observations of the dynamic cellular events in the embryonic cortex led to the radial unit hypothesis which postulates that the size of the cerebral cortex depends on the number of contributing radial units, which in turn depends on the number of founder cells (Rakic 1988, 1995b). According to this hypothesis, the number of radial columns determines the size of cortical surface, whereas the number of cells within the columns determines its thickness.

Kinetics and mode of cell proliferation

The size of the cortex is determined in the proliferative zones before cells migrate to the cortical plate. The total neuronal number in the cortex depends on several factors, including the number of founder cells, the time of onset of corticogenesis, the duration of the cell division cycle, the duration of the period of neurogenesis, the modes of cell division, the number of rounds of cell cycles and finally selective programmed cell death (apoptosis). Progress made in understanding specific contributions of some of these factors is reviewed elsewhere (Rakic & Kornack 2000, Takahashi et al 1997) and only a brief account is provided here.

In the mouse, with a gestation of 19 days, the cortex is generated in about one week, between embryonic day (E) 12 and term. In contrast, in the macaque monkey, with a 165-day gestation, all cortical neurons are generated within the two-month period, between E40 and E100. Finally, in the human, with a 40-week gestation, corticogenesis lasts more than three months, from the end of the sixth week (E42) to about the 20th week (Reviewed in Sidman & Rakic 1973, 1982). Before neurogenesis starts, the common progenitor (founder) cells in all three species are dividing symmetrically: each progenitor produces two additional progenitor cells during each mitotic cycle (Rakic 1988). Thus, during this phase the number of progenitor cells is doubling with each extra round of symmetrical divisions, resulting in an exponential increase in the size of the ventricular zone (Fig. 2A). As a result, a slight prolongation of this phase of telencephalic development of proliferation could be indirectly responsible for a significant surface enlargement of the cerebral cortex (Rakic 1995b).

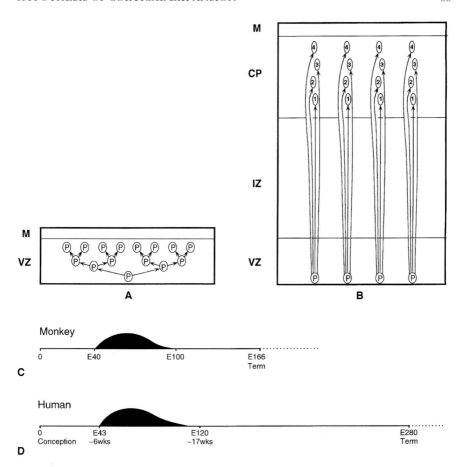

FIG. 2. (A) Schematic model of symmetrical cell divisions which predominate before the 40th embryonic day (E40). At this early embryonic age, the cerebral wall consists of only the ventricular zone (VZ), where all cells proliferate and the marginal zone (M), where some of them extend their radial processes. Symmetrical division produces two progenitors (P) during each cycle and causes rapid horizontal lateral spread. (B) Model of asymmetrical or stem division which becomes predominant in the monkey embryo after E40. During each asymmetrical division a progenitor (P) produces one postmitotic neuron which leaves the ventricular zone and another progenitor which remains within the proliferative zone and continues to divide. Postmitotic neurons migrate rapidly across the intermediate zone (IZ) and become arranged vertically in the cortical plate (CP) in reverse order of their arrival (1, 2, 3, 4). (C) Diagrammatic representation of the time of neuron origin in macaque monkey. The data are obtained from ^{3}H-thymidine autoradiographic analyses (Rakic 1974). (D) Estimate of the time of neuron origin in the human neocortex based on the number of mitotic figures within the ventricular zone, supravital DNA synthesis in slice preparations of fetal tissue and the presence of migrating neurons in the intermediate zone of the human fetal cerebrum (from Rakic 1995b).

After E12 in mouse, E40 in monkey and about E42 in human, some progenitor cells begin to produce neurons that leave the ventricular zone. Autoradiographic analysis of the kinetics of cell proliferation indicates that, after E12 in mouse and E40 in monkey, neuronal precursors begin to divide asymmetrically (Rakic 1988, 1995b, Caviness et al 1995). This mode of division, also known as 'stem cell' division, produces one daughter cell, which is permanently postmitotic, and another, which continues to divide. The postmitotic cell that will become a neuron detaches from the ventricular surface and begins to migrate toward the pial surface, eventually settling in the cortical plate (Fig. 2B). The other daughter cell remains attached to the surface of the cerebral ventricle by an endfoot and continues to generate additional pairs of unequal cells. This mode of cell division in the monkey fetus lasts 30 to 60 days, depending on the cortical area (Rakic 1988). Thus, the kinetics of proliferative activity in the ventricular zone can be divided into two broad phases: (1) the phase when the founder cells of the prospective cerebral cortex are generated by symmetrical divisions; and (2) the phase of neurogenesis that proceeds mainly by an asymmetrical mode of cell division and continues until the completion of corticogenesis in a given region. The duration of the first phase and length of cell cycle in each species determines the number of radial units in the cortex and, indirectly, the size of the cortical surface. In contrast, the duration of the second phase regulates the number of neurons within each ontogenetic column. The elimination of isochronously dividing cells by low doses of ionizing radiation at early and late stages of embryonic development supports this model. Specifically, irradiation of monkey embryo before E40 results in a decrease in cortical surface with little effect on its thickness, whereas irradiation after E40 deletes individual layers, and reduces cortical thickness without an overall decrease in total surface (Algan & Rakic 1997).

Although the molecular mechanisms underlying the switch between the two phases of cortical development remain unclear, it is likely that the transition may be triggered by the activation of regulatory gene(s) that control the mode of mitotic division in the ventricular zone. For the purpose of the proposed model of cortical expansion, it is important to underscore that the major change in the mode of cell division in the telencephalon is initiated prior to arrival of the input from the subcortical structures.

Role of apoptotic genes

Several lines of evidence indicate that programmed cell death (apoptosis) is a major contributing factor to the formation of the vertebrate brain. In the past, most research has been focused on histogenetic cell death occurring at late developmental stages where it is primarily involved in eliminating 'incorrect' axonal connections after neurons have attained their final positions (Oppenheim 1991, Rakic 1986, Cowan et al 1984). Recently, the use of methods that identify dying cells by exposing their fragmented nuclear DNA suggests that apoptosis in the ventricular zone is more significant than assumed (Blascke et al 1996). The possibility of studying this issue in mammals at the molecular level has dramatically increased after identification of regulatory genes that regulate programmed cell death (Metzstein et al 1998). Because biologically important molecules tend to be conserved during evolution (e.g. Wolpert 2000, this volume), we can examine the function of deleted genes during brain development. This can be illustrated by our recent work on the genes controlling apoptosis in the mouse embryo (Kuida et al 1996, 1998). The rate of apoptosis can be reduced by inactivating genes for caspase 3 and 9 that must be turned on for a cell to die. In mice lacking both copies of the gene, apoptosis is reduced in the cerebral ventricular zone at early stages, during production of the founder progenitor cells. The finding that is most relevant to the subject of the present workshop is that, in accord with the radial unit hypothesis, a larger-than-normal number of founder cells resulted in a cortical plate with an increased surface area and the formation of convolutions (Fig. 3). By this single gene mutation, a lyssencephalic mouse cortex was transformed into gyrencephalic cerebrum, which is usually a hallmark of larger brains, as if there was a recapitulation of evolution (reviewed in Haydar et al 1999). I should underscore that programmed cell death is not the only, or even the major, factor leading to the increase in cortical size during evolution. However, this example illustrates how protection of a small number of specific founder cells may significantly affect the final number of neurons generated during later stages of development. In addition, while larger numbers of generated cells in the mutant mice presented here result in formation of incipient cortical convolutions, the precise species-specific pattern of cerebral convolutions also depends on connectivity between cortical areas, as well as with subcortical targets (Goldman-Rakic & Rakic 1984, Rakic 1988, Van Essen 1997).

In this instance, the mutation resulting in more cortical neurons was not good for the organism—most of the homozygous mice died before birth. However, during evolution, over millions of years, numerous mutations that increase the number of founder cells by either changes in the kinetics of proliferation or cell death could occur, and at some point supernumerary cells may have formed functionally useful connections that have helped in the survival of the species. Although the developmental mechanisms underlying the natural occurrence of cortical gyri in other species and their specific placement and orientation remain largely unknown, several theories have been proposed, which are beyond the scope of this review (e.g. Richman et al 1975,

Wild-type   Casp-9 KO

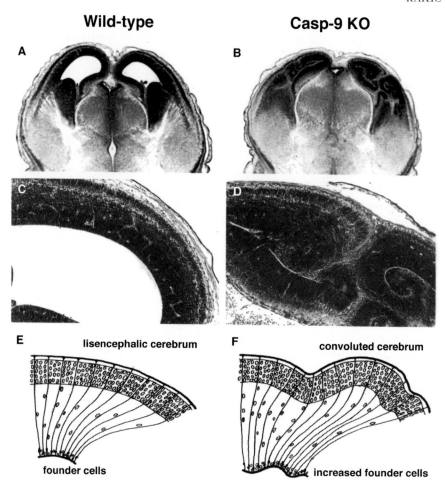

FIG. 3.   Caspase 9 (Casp-9) regulates the size of the neuroepithelial progenitor pool. Gene targeting of caspase 9 typically produces an expanded and convoluted cerebral cortex with an increased number of neurons (compare A, C in wild-type to B, D in mutants). (E, F) These abnormalities suggest that the caspase 9 null mutation primarily affects apoptosis of the founder cells in the ventricular zone. Adapted from Kuida et al (1998).

Welker 1990, Armstrong et al 1995, Van Essen 1997). This experiment addresses this issue and illustrates the remarkable power of molecular and developmental neurobiology. Here we used a gene identified in a roundworm that may help our understanding of the principles of cortical development that could be extrapolated to primate evolution.

Cortical expansion is a first step

To develop highly interconnected multiple functional areas that are essential for elaborate human cognition, the cerebral cortex first has to acquire a large number of neurons arranged as a continuous flat sheet. We have suggested that the evolutionary expansion of the neocortex in primates could be attributed to a change in the genetic mechanisms that control cell production in the ventricular zone and their allocation in the developing cortical plate (Rakic 1988, 1995b). As illustrated in Fig. 2, during the first phase, each round of symmetrical cell divisions could double the number of founder cells, whereas, during the second phase, when asymmetrical divisions begin to predominate, each additional mitotic cycle adds only a single neuron to a given ontogenetic column (Fig. 2A, B). Indeed, in the monkey, the first phase lasts four weeks longer than it does in the mouse so that, although the length of the cell cycle at the onset of neuron production is about twice as long than in mice (Kornack & Rakic 1998), the size of proliferative pool at comparable embryonic stages is much larger in monkey than in mouse. In contrast, a delay in onset of neurogenesis of only a few days can account for the difference on an order of magnitude of cortical expansion between monkey and human (Fig. 2C, D). If the length of cell cycle in these two Old World primates is comparable, a delay in the onset of the second phase that allows three to four extra rounds of mitosis would result in a 2^3- to 2^4-fold increase in founder cells that would generate an eight- to 16-fold larger number of columns, and, therefore, a proportionally larger cortical surface (Fig. 2C, D). In contrast, a 20-day delay in cessation of the second phase in human compared to monkey (E100 and E120) would add only about 10 more cells per ontogenetic column. Assuming that each column consists of about 100 neurons (Rakic 1988), such an addition would increase cortical thickness by only 10%. For example, this model does not take into account the possible changes in the proportion of symmetrical cell divisions during the course of the second phase, the growth in size of individual neurons, the contribution of glial cells and myelin, or the rate of cell death, all of which may also influence surface expansion to different extents in each species. Nevertheless, these developmental and structural differences are relatively minor in the Old World primates selected for comparison and an increase in the number of radial units is likely be the most prominent and decisive evolutionary factor.

According to the proposed model, the species-specific size of the cortex is determined at early stages by the pool of founder cells before corticogenesis starts, and before there is any input from the periphery. The evolutionary construction of the mammalian brain may require as many genes as were needed for other morphogenetic and metabolic functions in phyletic history (John & Miklos 1988). However, a small modification of a regulatory gene(s) may provide a first step in the evolutionary expansion of the neocortex, as presumably has occurred in other bodily systems (e.g. Medawar 1953). Therefore, the explanation for cortical expansion in mammalian evolution rests predominantly upon the process of heterochrony, whereby changes in the timing of developmental events, in this case kinetics of cell proliferation, increase the number of founder cells and consequently the surface of the cortical plate (Rakic 1995b, Kornack & Rakic 1998). Nevertheless, the enlargement of cortical surface area or even the differential expansion of the individual areas in themselves are not sufficient to account entirely for the elaboration of cortical connectivity that occurred during evolution. The additional issues, such as the creation of new cortical areas, are discussed by others at this symposium. However, the increase of cortical surface by the introduction of additional radial units, as well as the expansion and elaboration of cytoarchitectonic areas, provides an opportunity for creating novel input/target/output relationships with other structures that, if heritable, may be subject to natural selection. The new synaptic relationships resulting from these neuronal interactions may be adverse, neutral or may enhance capacity for behavioural adaptation. As pointed out by Jacob (1977) a new structural feature does not have to be optimal, but must be 'good enough' to provide a survival advantage for the species.

Acknowledgements

This work was supported by the US Public Health Service. I am grateful to the present and past members of my laboratory for their valuable contributions and interesting dissuasions.

References

Algan O, Rakic P 1997 Radiation-induced, lamina-specific deletion of neurons in the primate visual cortex. J Comp Neurol 381:335–352

Armstrong E, Schleicher A, Omram H, Curtis M, Zilles K 1995 The ontogeny of human gyrification. Cereb Cortex 5:56–63

Blaschke AJ, Staley K, Chun J 1996 Widespread programmed cell death in proliferative and postmitotic regions of the fetal cerebral cortex. Development 122:1165–1174

Blinkov SM, Glezer IJ 1968 The human brain in figures and tables, a quantitative handbook. Plenum Press, New York

Cameron RS, Rakic P 1991 Glial cell lineage in the cerebral cortex: a review and synthesis. Glia 4:124–137

Caviness VS Jr, Takahashi T, Nowakowski RS 1995 Numbers, time and neocortical neuronogenesis: a general developmental and evolutionary model. Trends Neurosci 18: 379–383

Cowan WM, Fawcett JW, O'Leary DD, Stanfield BB 1984 Regressive events in neurogenesis. Science 225:1258–1265

Goldman-Rakic PS. Rakic P 1984 Experimental modification of gyral patterns. In: Geschwind N, Galaburda AM (eds) Cerebral dominance: the biological foundation. Harvard University Press, Cambridge, MA, p 179–192

Haydar TF, Kuan C, Flavell RA, Rakic P 1999 The role of cell death in regulating the size and shape of the mammalian forebrain. Cereb Cortex 9:621–626

Jacob F 1977 Evolution and tinkering. Science 196:1161–1166

John B, Miklos GL 1988 The eukaryote genome in development and evolution. Allen & Unwin, London

Komuro H, Rakic P 1998 Orchestration of neuronal migration by the activity of ion channels, neurotransmitter receptors and intracellular Ca^{2+} fluctuations. J Neurobiol 37:110–130

Kornack DR, Rakic P 1995 Radial and horizontal deployment of clonally related cells in the primate neocortex: relationship to distinct mitotic lineages. Neuron 15:311–321

Kornack DR, Rakic P 1998 Changes in cell-cycle kinetics during the development and evolution of primate neocortex. Proc Natl Acad Sci USA 95:1242–1246

Kuida K, Zheng TS, Na S et al 1996 Decreased apoptosis in the brain and premature lethality in CPP32-deficient mice. Nature 384:368–372

Kuida K, Haydar T, Kuan C et al 1998 Reduced apoptosis and cytochrome c-mediated caspase activation in mice lacking caspase 9. Cell 94:325–337

Lois C, García-Verdugo JM, Alvarez-Buylla A 1996 Chain migration of neuronal precursors. Science 271:978–981

Luskin MB, Pearlman AL, Sanes JR 1988 Cell lineage in the cerebral cortex of the mouse studied *in vivo* and *in vitro* with a recombinant retrovirus. Neuron 1:635–647

Medawar PB 1953 Some immunological and endocrinological problems raised by the evolution of vivparity in vertebrates. Symp Soc Exp Biol 7:320–338

Metzstein MM, Stanfield GM, Horvitz HR 1998 Genetics of programmed cell death in *C. elegans*: past, present and future. Trends Genet 14:410–416

Misson JP, Austin CP, Takahashi T, Cepko CL, Caviness VS Jr 1991 The alignment of migrating neural cells in relation to the murine neopallial radial glial fiber system. Cereb Cortex 1:221–229

Mountcastle VB 1997 The columnar organization of the neocortex. Brain 120:701–722

Oppenheim RW 1991 Cell death during development of the nervous system. Annu Rev Neurosci 14:453–501

Pearlman AL, Faust PL, Hatten ME, Brunstrom JE 1998 New directions for neuronal migration. Curr Opin Neurobiol 8:45–54

Rakic P 1972 Mode of cell migration to the superficial layers of fetal monkey neocortex. J Comp Neurol 145:61–83

Rakic P 1974 Neurons in rhesus monkey visual cortex: systematic relation between time of origin and eventual disposition. Science 183:425–427

Rakic P 1978 Neuronal migration and contact guidance in primate telencephalon. Postgrad Med J 54:25–40

Rakic P 1981 Neuron–glial interaction during brain development. Trends Neurosci 4:184–187

Rakic P 1986 Mechanism of ocular dominance segregation in the lateral geniculate nucleus: competitive elimination hypothesis. Trends Neurosci 9:11–15

Rakic P 1988 Specification of cerebral cortical areas. Science 241:170–176

Rakic P 1990 Principles of neuronal cell migration. Experientia 46:882–891

Rakic P 1995a Radial versus tangential migration of neuronal clones in the developing cerebral cortex. Proc Natl Acad Sci USA 92:11323–11327

Rakic P 1995b A small step for the cell, a giant leap for mankind: a hypothesis of neocortical expansion during evolution. Trends Neurosci 18:383–388

Rakic P 1997 Intra and extracellular control of neuronal migration: relevance to cortical malformations. In: Galaburda AM, Christen Y (eds) Normal and abnormal development of cortex. Springer-Verlag, Berlin (Res Perspect Neurosci) p 81–89

Rakic P, Kornack DR 2000 Neocortical expansion and elaboration during primate evolution. a view from neuroembryology. In: Gibson K, Falk D (ed) Evolutionary anatomy of primate cerebral cortex. Cambridge University Press, Cambridge, in press

Rakic P, Stensaas LJ, Sayre EP, Sidman RL 1974 Computer-aided three-dimensional reconstruction and quantitative analysis of cells from serial electronmicroscopic montages of foetal monkey brain. Nature 250:31–34

Rakic P, Cameron RS, Komuro H 1994 Recognition, adhesion, transmembrane signaling, and cell motility in guided neuronal migration. Curr Opin Neurobiol 4:63–69

Richman DP, Steward RM, Hutchinson JW, Caviness VS Jr 1975 Mechanical model of brain convolutional development. Science 189:18–21

Rubenstein JLR, Anderson S, Shi L, Miyashita-Lin E, Bulfone A, Hevner R 1999 Genetic control of cortical regionalization and connectivity. Cereb Cortex 9:524–532

Sidman RL, Rakic P 1973 Neuronal migration with special reference to developing human brain: a review. Brain Res 62:1–35

Sidman RL, Rakic P 1982 Development of the human central nervous system. In: Haymaker W, Adams RD (eds) Histology and histopathology of the nervous system. CC Thomas, Springfield, IL, p 3–145

Takahashi T, Nowakowski RS, Caviness VS Jr 1997 The mathematics of neocortical neuronogenesis. Dev Neurosci 19:17–22

Tan SS, Kalloniatis M, Sturm K, Tam PPL, Reese BE, Faulkner-Jones B 1998 Separate progenitors for radial and tangential cell dispersion during development of the cerebral neocortex. Neuron 21:295–304

Van Essen DC 1997 A tension-based theory of morphogenesis and compact wiring in the central nervous system. Nature 385:313–318

Ware ML, Travazoie SF, Reid CB, Walsh CA 1999 Coexistence of widespread clones and large radial clones in early embryonic ferret cortex. Cereb Cortex 9:636–645

Welker W 1990 Why does cerebral cortex fissure and fold? In: Jones EG, Peters A (eds) Cerebral cortex, vol 8b: Comparative structure and evolution of cerebral cortex, part III. Plenum Press, New York, p 3–136

Wolpert L 2000 What is evolutionary developmental biology? In: Evolutionary developmental biology of the cerebral cortex. Wiley, Chichester (Novartis Found Symp 228) p 1–14

DISCUSSION

Wolpert: When you knocked out the caspase, you observed almost a doubling of cortical size. This means that half of the stem cells should undergo cell death, but this did not seem to be the case from your TUNEL stain results.

Rakic: The surface of the cortex in the mutants is visibly larger, but it is not double the size. Furthermore, the sections stained with the TUNEL method that you refer to in your question, illustrates the cerebral wall at the late stage of corticogenesis, when neurons destined for the cortex are being generated. At this stage we see relatively few TUNEL-positive cells. The deletion of caspases 3 and 9 reduces the rate of programmed cell death (apoptosis) in the proliferative zones during the early stages of embryogenesis of the forebrain when cortical founder cells are being formed; therefore, before the onset of corticiogenesis. This does not mean that some cells are not spared from dying also at the late stages. However, sparing of a relatively small number of progenitors at an early embryonic stage may have a large consequence for the final size of the cerebral cortex: a few founder cells spared from death at the early stages might produce a large number of cortical neurons if they continue to divide by symmetrical divisions. Our working hypothesis that explains the actual results obtained is that normally some founder cells die early in the ventricular zone and that, by prevention, their death in the mutants results is a larger cortex than in the wild-type controls.

Hunt: The TUNEL technique applied at the late stages only gives a brief snapshot of the overall process.

Wolpert: So, in other words that snapshot tells you nothing about the overall amount of cell death.

Rakic: That is correct. The TUNEL method at that stage does not tell what may have happened in the proliferative zone before the onset of corticogenesis. Our results in knockout mice illustrate how a relatively small number of dying founder cells can have a large effect on the final outcome.

Pettigrew: There are knockout mice of genes involved in the timing of proliferation. In these mice, you observe the same sort of paradox that Lewis Wolpert is pointing out. The brain isn't as large as you would expect it to be. One hypothesis that has been put forward is that there are interrelationships between the caspase genes and other genes such as the cyclins. Therefore, interactions between the developmental programmes that are generating cells and programmes that are killing them could give rise to these sorts of discrepancies.

Rakic: I agree. The final outcome from the proliferative zones depends on the balance between many factors: duration of cell cycle, mode of cell division, duration of neurogenesis, etc. For example, some genes regulate pattern and distribution of dying cells in the proliferate zones. We have recently found that some members of the JNK family of protein kinases play a role in the regulation of cell death pattern in the proliferate zone. In the double *JNK1/2* knockout mice, apoptotic cells are not distributed at the appropriate positions within the proliferative epithelium and, as a consequence, the neural tube does not close and a larger number of cells die in the prospective forebrain (Kuan et al 1999).

Levitt: Jerald Chun and colleagues have also shown, using a different technique, that there is significantly more cell death than can be demonstrated by the TUNEL staining technique (Blaschke et al 1996, 1998).

Rakic: Yes. However, he was studying a different stage of cortical development, i.e. the stage when neurons destined for the cortex are being formed. In contrast, we are discussing here founder cells, the progenitors that divide by symmetrical division before the onset of neurogenesis.

Papalopulu: How do you know that the cells that are not dying are progenitor cells?

Rakic: We see fewer pyknotic cells and more progenitors in the ventricular zone of the heterozygous mutants. This can be seen clearly on routine histological material. Furthermore, we have observed the effect of this mutation, which is a larger number of neurons. So all of this fits together rather convincingly.

Rubenstein: Did you do a TUNEL assay in the mutant to look at whether there are any changes in the number of TUNEL-positive cells?

Rakic: Yes, we did; but we encountered variations in the number of cells stained, depending on the procedure used, so we did not use this method for quantification. We found that electron microscopy is the most reliable method for identification of apoptotic cells. However, this method is also not suitable for quantification. Perhaps the most practical method for surveying large areas is plotting pyknotic cells in 1 μm-thick sections.

Karten: But in your presentation you showed a slide of TUNEL-stained cells, and you pointed out that cell death was sparse. This supported your argument that the TUNEL method produces an overestimation of cell death.

Rakic: I am not suggesting that the TUNEL method is not useful. My point was that sparing cell death at late stages of corticogenesis couldn't fully explain our results. In contrast, only a few cells spared at the early stage could have the large effect that we observed.

Molnár: You said that the *CPP322* (caspase 3) mutant survived to adulthood. What is the cortical regionalization in these animals? Do they have barrel fields, for example?

Rakic: Only a few of these mutants survive. Many develop hydrocephalus and die before reaching adulthood. The ones that do survive, however, have barrel fields and an apparently normal spinal cord.

O'Leary: From the perspective of building a larger cortex, do you observe the same phenotypic ratios in the mutant and the wild-type?

Rakic: So far, we have only looked at the distribution and ratio of parvabumin that stains interneurons and found about 30% of immunopositive cells in the cortex as well as in the ectopic subcortical regions, which is about the same as in the normal controls.

Reiner: I wanted to ask a sort of chicken-and-egg question about the evolution of cortex. You said that the driving event for the evolution of cortex is expansion of cortex via increased numbers of radiocortical units, and you seemed to say that a secondary event is the invasion of those units by thalamic afferents. Does cortical expansion occur first, followed by a period of evolution when there's no thalamic input; is the thalamic input automatically drawn to a neocortical area; or does the thalamus drive the expansion of the cortex?

Rakic: Expansion of the cortex during evolution requires there to be a mechanism to generate a larger number of cells. Several lines of evidence indicate that the species-specific size of the cortex is set at early stages, before thalamic input arrives at the cortical plate. However, a larger number of neurons in the cortex is not sufficient or even necessarily beneficial for the organism since in most instances these malformations are lethal. For example, a mutation generating a larger cortex, but not a larger thalamus to accommodate this expansion, may not be useful. At some point in evolution, however, a mutation may have occurred that enabled the formation of functionally useful connections. This animal would then survive, reproduce and propagate that trait. Some people have asked me if the mouse with more neurons is behaviourally smarter than its littermates. In this case, apparently, more cells are not good for the animals, as they die. To produce a smarter mouse we would need a million years to test many combinations by trial and error, as it may have happened during evolution!

Herrup: When you go back in developmental time, when is the first point where you can reliably distinguish between your knockouts and the wild-type? In other words, how soon does cell death make a substantial difference in the mouse?

Rakic: Before the onset of cortex formation. In this mutant, the rhombencephalic vesicle doesn't close completely. The reason for this failure may be due to cell death, which is required for neurulation (Kuan et al 1999).

References

Blaschke A J, Staley K, Chun J 1996 Widespread programmed cell death in proliferative and postmitotic regions of the fetal cerebral cortex. Development 122:1165–1174

Blaschke A J, Weiner J A, Chun J 1998 Programmed cell death is a universal feature of embryonic and postnatal neuroproliferative regions throughout the central nervous system. J Comp Neurol 396:39–50

Kuan, C-Y, Yang DD, Samanta-Roy DR, Davis R J, Rakic P, Flavell A 1999 The Jnk1 and Jnk2 protein kinases are required for regional specific apoptosis during early brain development. Neuron 22:667–676

General discussion I

Evolution of cell populations

Karten: I would like to mention some of my own work on the issue of how brains differ. The essential question is, what is it that has evolved in the forebrain? The most striking features that have evolved in the case of the cortex are the lamination and the specificities of both the inputs and outputs characteristic of each layer of cortex.

In the rat brain, there are distinctive regions across the different areas of cortex. In stark contrast, in a non-mammalian brain there is just a thin cortical zone. For many years the notion was that mammalian cortex arises by *de novo* neurogenesis within the dorsal pallium area, and that most pallial areas in non-mammalian amniotes are like basal ganglia and coalesce into a set of basal nuclei. In order to understand what tinkering may have occurred and whether it is conservative, we have to compare the neurons in the cortex of mammals and non-mammals. Alternatively, we could look at this from a behaviouralist's point of view. Birds are capable of extraordinary feats of auditory, visual and somatosensory discrimination. Those are properties that are endowed to us in many ways by our cortex. What structures do non-mammals use instead?

The original notion was that it somehow involved the basal ganglia. Over many years, we did many sets of studies which showed that only part of the large nuclear region in the bird telencephalon corresponds to the mammalian caudate putamen and globus pallidus, and the rest of this region, which is called the dorsal ventricular ridge (DVR), is something else. What is the nature of this 'something else'? Instead of looking at this region as a global field, we recognized the need to look at the neurons, their identities, their inputs, outputs, transmitters and so forth. The constituent populations are one thing, and lamination is another. One of the sources of confusion in regard to studies of evolution is that we assume that lamination and cortical cell populations must co-evolve. However, they evolve independently, just as the adrenal cortex and the medulla evolve independently in vertebrates. They arise as independent organs, though in most animals they migrate to form a single 'organ'. They change in their operations, but not in their fundamental resulting qualities, as a consequence of evolutionary modification.

Does the same thing occur in the brain? Is lamination an independent event that modifies some of the consequences but not the fundamental identities? The major players are the pallial zone, and bulging into the ventricle, the basal region and the dorsal ridge. For many years the notion was that anything that sticks into the ventricle has to be basal ganglia. But now we know that the dorsal ridge is more related to the pallium than was suggested by earlier studies. If this region is related to the pallium, what is the nature of its neurons? We need to address this question in order to understand how it evolved. There is only a small area that corresponds to basal ganglia, so what is the nature of the rest of this dorsal ridge? One hypothesis is that cortex did not evolve *de novo* with the appearance of mammals. It can be analysed with respect to two properties: the constituent populations and the laminar appositions. What are the general themes of cortical organization? In a simplified model, we can talk of cortex as if it consists of three major groups of neurons: (1) those that receive a thalamic input and arborize intrinsically within the cortex; (2) those that are purely interneurons; and (3) those that give rise to descending efferents. Sets of these neuronal groups are specific to different sensory modalities. Are there similar structures in non-mammals? If we start to pursue this by starting at the far periphery, we find that there is a specific auditory pathway which comes up from the inferior colliculus and goes first to the thalamus and then to the telencephalon. It projects into the zone that we thought for so many years was a basal ganglia-like zone, i.e. the DVR. We now recognize that there are highly specific tonotopic bands, in this case called L2A, that project to other neurons which in turn go to a discrete group of cells that provide the long descending outputs. The similarities between that arrangement and the arrangement of the simplified cortex are clear, i.e. they are thalamic recipient neurons, interneurons and the descending telencephalic efferents.

Let us compare the situation in the auditory pathway of birds and mammals. In mammals we have thalamic inputs to layer IV, interneurons nominally in layers II/III projecting onto layers V/VI and then descending efferents to the thalamus and brainstem. In birds the groups are not necessarily laminated, but they have the same physiological properties. They also have many, although not all, similar three-dimensional morphological properties, ion channels and transmitters; and the destruction of many of these fields produces similar deficits in behaviour as would the destruction of the specific sensory cortices in mammals. But they are not laminated in an obvious way. One way to think of this is to give an example of one of the visual pathways, of which there are several in the forebrain. The thalamic recipient interneurons and output neurons are in separate regions, whereas in the mammalian cortex they are in a stratified configuration.

There are a number of laminated structures in nervous systems — e.g. the olfactory bulb, the cortex, the retina and the cerebellum — and there are even structures at brainstem and spinal cord levels that are highly laminated. A dramatic example of this is the facial and vagal lobes of fish. Fish have all sorts of specialized laminated structures in their brainstem. These depend upon modality specificity and how they have adapted to their environment. Catfish, for example, have a wonderful gustatory system with barbels that are coated with taste buds. They project to a highly somatotopic or gustatopic structure in the brain, the facial lobe. They also have a gustatory system inside the mouth that projects via the vagal system to the vagal lobe. In contrast, the goldfish, which has the same sets of neurons in the brainstem (i.e. the facial and vagal lobes) has a large organ in the roof of its mouth called the pallital organ. This is covered with taste buds, and does precise manipulations inside the mouth, in contrast to the catfish who chooses its meal externally. In both goldfish and catfish these lobes are, to all intents and purposes, elegant laminated structures.

Lamination is an independent event in evolution. What are the genetic changes involved in generating laminated structures? What are the molecular mechanisms underlying cell migration, elaboration and differentiation? Lamination is not new, and it can be variable. In the case of the forebrain, there is a basal ridge, which is mostly but not exclusively basal ganglia, and a number of other bulges. These bulges can mostly be called prosomeres. We don't like this term because it has a totally different meaning within cell biology. There is a serious question as to whether we should call these things neuromeres within the telencephalon. Holmgren (1925), an inspired Swedish developmental biologist, and his student Kállén, published a series of papers in which he attempted to revive the concepts of neuromeres at the brainstem level. He admitted that the forebrain was difficult because it didn't conform. What we may imagine is that these constituents contain various populations of cells, some of which may migrate to different regions of the laminated zone to make the different kinds of cortices. The question that emerges from this is, is radial migration an essential component of the mammalian isocortex, or are there other ways of making cortex? If there are early prosomeres, what is their fate? Are the constituents of these prosomeres translocated and incorporated to make a cortex?

In 1991 I restated the observations of Stensaas & Gilson (1972) concerning the zone that they called the subventricular zone. This may be a different use of the term 'subventricular', but they pointed out that the subventricular zone sits at the dorsal lateral margin of the lateral portion of the lateral ganglionic eminence (LGE). Data from John Rubenstein's lab and John Parnavelas' lab, among others, suggest that cortex is not made up only of cells with pallial origin, a point that was suggested by a number of people over the years. In birds and reptiles, cells migrate locally to form the DVR. In mammals, at least some of the cortical areas are derived by this lateral migration.

Therefore, the constituents that make up the cortex have evolved independently. There is abundant evidence from birds and reptiles. We don't know about early reptiles and amphibians, because modern amphibia may not be adequate representatives of the stem tetrapods. We can't do this circuitry analysis in fossilized brains. From this point of view, one might ask, what are the constancies and what are the mechanisms of change involved?

Parnavelas: The area you have pointed to is indeed the subventricular zone. In the rat, this zone appears adjacent to the lateral ventricle at the time of birth and persists in that position throughout the life of the animal. Placement of tracer in the subventricular zone of postnatal rats labels glial cells that migrate to the cerebral cortex and neurons destined for the olfactory bulb.

Karten: All the data we have suggest that all the neurons forming in the DVR form well before the vast majority of the cortical neurons. They are different areas, so perhaps this is the answer to the disagreement. Many of the neurons that make up cortex are born at embryonic day 9.5–11.5. Particularly in the lateral cortex, there is a striking gradient of temporal to medial cortex.

Rubenstein: In our view, the DVR is a cortical structure, and is not relevant to the studies on the generation of interneuron precursors in the basal ganglia.

Purves: To strengthen your argument in the case of birds, you might include data from the bird-song system in finches and canaries. Can you incorporate the bird-song system into your evolutionary argument?

Karten: We can for some of the major constituents. The early take-off point for that was from field L, i.e. the auditory field. Bonke et al (1979) and Kelley & Nottebohm (1979) identified that there were outputs to other areas, including what was called HVC, which now turns out to be an erroneous name. This is comparable to some of the areas seen in non-song birds. The RA area, which is a large descending pre-motor nucleus, is also within the DVR, but caudolaterally within the archistratum. Most of these points fit my hypothesis, but do not prove it. Area X turns out to be a difficult area, but HVC and RA are similar to cortical neurons in terms of being derivatives of the dorsal ventricular region. We know of no comparable systems that parallel the highly specific neurons at field L, HVC or RA within a claustrum.

Puelles: You simplified matters somewhat because you only mentioned the basal ganglia and cortex. The claustrum, amygdala and the endopyriform nucleus are also classically known to be pallial in origin and nuclear in structure, and they are intercalated between the cortex and the subpallium. It is therefore possible that some of the formations present in the sauropsidian DVR may be neither basal ganglia nor cortical primordia, but primordia of these other structures that you did not mention. In any case, we not only need to ascertain where is the cortex of birds, but also where is its claustroamygdaloid complex. Only a complete explanation will be satisfactory.

Karten: There are also other areas in birds and mammals that have not been studied extensively, and we don't yet know what they do.

Hunt: I am surprised that there is any need for cortex in birds. They have a small region called the hippocampal cortex, which varies in size and is important in birds that store food. Is it not problematic that there is any cortex, given that the neostriatum is thought to do the same kind of analysis as cortex in mammals?

Karten: There is a structure called a Wulst in birds, which is almost a classical visual or striate cortex. It is laminated, it has retinotopic inputs and it receives inputs from a structure much like the dorsal geniculate nucleus. In owls and animals with bifrontal eyes, there are stereo units and there is disparity detection. My discovery of the visual nature of the Wulst as a 'cortex' brings up a different issue, which is that within the telencephalon different cortical areas may form by different mechanisms. The cortex has six layers, and we are confronted with a vast area. We search for whatever may provide a unifying quality and then assume that this unifying quality also represents a unifying embryogenesis. My guess is that there is more than one mechanism for making cortex.

Butler: I've argued before (Butler 1994a,b, 1995) that there are two fundamental divisions in the cortex. In mammals, the striate cortex, somatosensory cortex and frontal cortex are derived from a separate division of the pallium than the lateral temporal neocortex. There are also two corresponding, fundamental divisions in the dorsal thalamus that have different connections. One division, called the collothalamus, receives input from the midbrain roof and projects to temporal and extrastriate cortical regions in mammals and to the anterior DVR in birds. The lemnothalamus derives from the rostral part of the dorsal thalamus. In mammals, the latter division includes the dorsal lateral geniculate nucleus, which migrates caudally, the ventral nuclear group, and the medial and anterior groups. It is the lemnothalamic nuclei in birds that project to the Wulst. Due to the more dramatic cytoarchitectural differences between the Wulst and the DVR in birds, there is a more obvious split between these pallial divisions in birds than in mammals, and differences in embryogenesis produce these different phenotypes.

Rakic: You mentioned that there is more than one way to provide the cortex with a new kind of neuron. If some cells originate from the DVR in a common ancestor, this does not necessarily contradict the radial unit hypothesis of cortical expansion. These precursors could be incorporated within the ventricular zone and form a cell population that eventually migrates radially. Radial migration has to explain the increased size of the cortex, whereas you are talking about the origin of founder cells that form the cortex, and whether they have common ancestors in birds and mammals. This is an artificial distinction because cells could have the same origin but still generate diversity.

Karten: It is not an artificial distinction; it is a valid distinction. The cells may either be translocated or transformed, i.e. displaced, in some way. One of the notions I've entertained is that the anlage of populations migrates and is incorporated into the ventricular proliferative zone. However, we do not have direct evidence for this. The difference between what we are able to discern when we have an organism of 330 neurons versus an organism with 10^{13} neurons, is different, in the sense that we can generate precise fate maps for the former. One of the major problems is that we cannot do a reverse time-lapse experiment, i.e. we cannot tag a neuron and run the clock backwards. What we really want to do in fate-mapping experiments is not label something early and then try to figure out which cells are labelled; we want to look at a particular neuron in the adult and find out where it came from.

Puelles: I don't see why one approach is better than the other. We need all the evidence we can get. At present, there is no evidence for the migration you postulate in your scheme, i.e. that cells produced at the LGE and sharing connectivity properties with given populations of the avian DVR migrate into the diverse cortical targets postulated in your hypothesis. The panorama has changed a bit in recent years, since the lateralmost edge of the classical LGE has been repeatedly shown not to express subpallial gene markers (i.e. the *distalless* gene family), so that morphology has here been corrected by molecular biology in the sense that this locus must be concluded to be a portion of the pallium and lies actually outside the molecularly defined LGE. Now, in this same position just external to the subpallium is where lack of *Emx1* expression and expression of *Tbr1* in the mantle and *Pax6* in the ventricular zone identifies an histogenetic field that clearly is strictly comparable to the avian neostriatum (Puelles et al 1999, 2000; see also Fernández et al 1998). The latter contains the thalamo-recipient cell populations which you identify a priori as potential cortical neurons. According to this scenario, what you have in fact predicted and not yet proven is that cells arising in this distinct pallial area of the ventricular zone lateral to the LGE proper migrate to the layer IV or other layers of the associative visual cortex and auditory cortex in mammals. Evidence from John Rubenstein's laboratory (Anderson et al 1997) suggests that numerous cortical interneurons indeed arise from the LGE proper. However, this is molecularly subpallial, so that these cells can be only compared to avian subpallial cells and not to the DVR neostriatal populations. Moreover, we have evidence from fate mapping studies in the chicken that many avian DVR and cortical interneurons also have a subpallial origin, so that such a tangential migration is not basically different in birds and mammals.

Karten: Which molecular markers are you using? Emx1?

Puelles: No. We are using quail–chick homotopic grafts and several antibodies to distinguish quail cells, neurons, glia and γ-aminobutyric acid (GABA)ergic interneurons (Cobos et al 2000). There is also evidence of *Dlx*-positive cells migrating out of the subpallium proper and dispersing within the pallium. Such massive movements are not targeted to specific areas in the pallium. In contrast, the fate mapping data of Striedter et al (1998) on the chick neostriatum show no tangential dispersion at all into the overlying parts of the pallium.

Karten: My reading of that data doesn't suggest this at all. If we look at the expression of *Tbr1*, we find that it is expressed throughout the entire DVR. Therefore, *Tbr1* does not give us sufficient specificity. I agree that there is no direct evidence in favour of my hypothesis, but there is also no evidence that discounts it. We also have to address why there is such a dramatic level of conservatism of neuronal circuits, transmitters and many other morphological properties.

References

Anderson SA, Eisenstat D, Shi L, Rubenstein JLR 1997 Interneuron migration from basal forebrain: dependence on *Dlx* genes. Science 278:474–476

Bonke BA, Bonke D, Scheich H 1979 Connectivity of the auditory forebrain nuclei in the guinea fowl (*Namida meleagris*). Cell Tissue Res 200:101–121

Butler AB 1994a The evolution of the dorsal thalamus of jawed vertebrates, including mammals: cladistic analysis and a new hypothesis. Brain Res Brain Res Rev 19:29–65

Butler AB 1994b The evolution of the dorsal pallium in the telencephalon of amniotes: cladistic analysis and a new hypothesis. Brain Res Brain Res Rev 19:66–101

Butler AB 1995 The dorsal thalamus of jawed vertebrates: a comparative viewpoint. Brain Behav Evol 46:209–223

Cobos I, Puelles L, Rubenstein JLR, Martinez S 2000 Fate map of the chicken rostral neural plate at stages 7–8, using quailchick homotopic grafts. in prep

Fernández AS, Pieau C, Repérant J, Boncinelli E, Wassef M 1998 Expression of the *Emx-1* and *Dlx-1* homeobox genes define three molecularly distinct domains in the telencephalon of mouse, chick, turtle and frog embryos: implications for the evolution of telencephalic subdivisions in amniotes. Development 125:2099–2111

Holmgren N 1925 Points of view concerning forebrain morphology in higher vertebrates. Acta Zool Stock 6:413–477

Kelley DA, Nottebohm F 1979 Projections of a telencephalic auditory nucleus — Field L — in the canary. J Comp Neurol 183:455–470

Puelles L, Kuwana E, Puelles E, Rubenstein JLR 1999 Comparison of the mammalian and avian telencephalon from the perspective of gene expression data. Eur J Morphol 37:139–150

Puelles L, Kuwana E, Puelles E et al 2000 Pallial and subpallial derivatives in the embryonic chick and mouse telencephalon. submitted

Stensaas LJ, Gilson BC 1972 Ependymal and subependymal cells of the caudato-pallial junction in the lateral ventricle of the neonatal rabbit. Z Zellforsch Mikrosk Anat 132:297–322

Striedter GF, Marchant TA, Beydler S 1998 The 'neostriatum' develops as part of the lateral pallium in birds. J Neurosci 18:5839–5849

Genetic control of regional identity in the developing vertebrate forebrain

Edoardo Boncinelli, Antonello Mallamaci and Luca Muzio

DIBIT, Scientific Institute San Raffaele, Via Olgettina 58, Milan, Italy

Abstract. In the past we isolated and characterized a number of vertebrate homeobox genes expressed in the developing brain. In particular, *Emx1* and *Emx2* are expressed in the developing forebrain of mouse embryos, in a region including the presumptive cerebral cortex. In the developing cerebral cortex, *Emx1* is expressed in most neuroblasts and neurons at all stages of development, whereas *Emx2* expression is restricted to proliferating neuroblasts of the so-called ventricular zone and to Cajal-Retzius cells, but is undetectable in most postmitotic cortical neurons. It is conceivable to hypothesize that *Emx2* plays a role in the control of proliferation of cortical neuroblasts and in the regulation of their subsequent migration. This latter process has been recently analysed in some detail in null mutant mice. The expression of these and other genes has also been analysed in the developing brain of different species of vertebrates. Homologies between forebrain subdivisions have been proposed based on the conservation and divergence of gene expression patterns.

2000 Evolutionary developmental biology of the cerebral cortex. Wiley, Chichester (Novartis Foundation Symposium 228) p 53–66

A number of regulatory genes have been isolated and characterized that play a role in the establishment and maintenance of the identity of anterior brain regions (Shimamura et al 1997, for review). Some of them are homeobox genes belonging to various gene families (Boncinelli 1999). Among them, the two genes belonging to the *Emx* family, originally found in mouse and in man (Simeone et al 1992a,b) and subsequently isolated in chicken (A. Mallamaci, unpublished results 1998), frog (Pannese et al 1998), and fish (Morita et al 1995, Patarnello et al 1997), occupy a particular position due to their restricted expression in dorsal forebrain, whereas the four genes of the *Dlx* family (Boncinelli 1994, for review) characterize specific regions of the developing ventral forebrain.

At embryonic day (E) 10 of mouse development the *Emx1* expression domain coincides with dorsal telencephalon. *Emx2* is expressed in dorsal and ventral neuroectoderm of the presumptive forebrain with a posterior boundary within the roof of the diencephalon. In the developing brain of embryos of this stage two homeobox genes of the *Otx* family are also expressed (Simeone et al 1992b).

The *Otx1* expression domain contains the *Emx2* domain and covers a continuous region including part of the telencephalon, the diencephalon and the mesencephalon. Finally, the *Otx2* expression domain contains the *Otx1* domain and covers the entire forebrain and midbrain with a posterior expression boundary coinciding with the boundary between the developing midbrain and hindbrain. Thus, the E10 brain shows a pattern of nested expression domains of the four genes in brain regions defining an embryonic rostral, or pre-isthmic, brain as opposed to hindbrain and spinal cord. The first appearance of the products of the four genes is also sequential during development: *Otx2* is expressed at least as early as at E5.5 in the implanted blastocyst (Simeone at al 1993), followed by *Otx1* and *Emx2* at E8–8.5 and finally by *Emx1* at E9.5. Here, we report on recent data concerning *Emx1* (and *Dlx1*) expression in the developing brain of various vertebrates (Fernández et al 1998) and the suggested role of *Emx2* in the control of the migration of cortical neurons (Mallamaci et al 1998, 2000).

Homologous expression patterns in the developing vertebrate forebrain

A number of features of brain organization have been conserved in vertebrate evolution, in particular in midbrain and hindbrain. Conversely, the organization of the forebrain is considerably more divergent and the homology of the various telencephalic subdivisions in mammals, birds and reptiles is still debated. The telencephalon of mammals is comprised of two major subdivisions: the cerebral cortex on the dorsal aspect and the basal ganglia located in deep ventral regions. The telencephalon of birds and reptiles is subdivided in three major domains: a dorsal cortical-like pallium, a subpallial formation bulging into the lateral ventricle, termed dorsal ventricular ridge (DVR), and a basal striatal domain. The relationship of dorsal and basal forebrain structures in birds, reptiles and mammals has been a subject of debate and is still controversial (reviewed in Northcutt & Kaas 1995).

A problem in its own right is the origin of the six-layered neocortex, or isocortex, of mammals. It has been claimed that it may derive from an ancestral dorsal pallium related to that present in extant amphibians (Northcutt & Kaas 1995). Alternatively, it has been proposed that the lateral portion of the neocortex and the anterior portion of the DVR (ADVR) share a common origin. In contrast, the posterior basal portion of the DVR (BDVR) shows histological and functional features that resemble those of the amygdala (reviewed in Striedter 1997).

We compared the expression of various genes, including *Emx1* and *Dlx1*, at a number of embryonic stages in a series of tetrapod species: *Mus*, *Gallus*, *Xenopus* and the turtle *Emys orbicularis* (Fernández et al 1998). Analysis of the expression domains of these genes suggested the presence of three main telencephalic subdivisions in all four species in the germinative neuroepithelium at the onset of neurogenesis. These subdivisions coincide with a pallial, an intermediate and a striatal domain, with the fate of the intermediate domain diverging between species at later stages of development. In the mouse this neuroepithelial region becomes rapidly vestigial and cannot be detected beyond E13.5. In chick and turtle this region increases in width during development and overlaps with the neuroepithelial lining of the growing DVR (Fernández et al 1998).

A relevant question concerns the fate of cell groups and neuronal structures deriving from the intermediate neuroepithelial domain in different species. A great part of the chick and turtle DVR is likely to derive from this territory, even if regions designated as DVR in birds and reptiles may not entirely coincide. To be more specific, the structures deriving from the lateral portion of the intermediate domain should constitute the reptilian DVR and the avian neostriatum, whereas the avian hyperstriatum should be considered as a pallial derivative. Based on gene expression and cytoarchitecture, additional discrete neuronal populations located in basal and medial telencephalon should be considered as derivatives of the early intermediate domain. In the mouse these are mostly located in the laterobasal part of the amygdala, as well as in the diagonal band and medial septum, whereas the corticomedial and central nuclei of the mouse amygdalar complex are likely derivatives of the dorsal neuroepithelial domain (Fernández et al 1998).

Even if it is highly debatable that comparison of gene expression patterns in different species constitutes *per se* a valid criterion for suggesting a common descent, it is possible on the basis of these studies to suggest homologies between the various forebrain subdivisions.

EMX2 and cortical migration

The expression of the two *Emx* genes in the developing cerebral cortex has been extensively studied (Simeone et al 1992a,b, Gulisano et al 1996, Briata et al 1996, Mallamaci et al 1998). *Emx1* is expressed in most neuroblasts and neurons at all stages of development, whereas *Emx2* expression is restricted to proliferating neuroblasts of the so-called ventricular zone and is undetectable in most postmitotic neurons. Very little is presently known about the role played by *Emx1* in the developing cortex, whereas some light is beginning to be cast on *Emx2* function.

Emx2 has been knocked-out in mice by homologous recombination in embryonic stem cells (Pellegrini et al 1996, Yoshida et al 1997). Homozygous null mutant mice die perinatally, probably because of the absence of kidneys. The neocortex of these mutant embryos is greatly reduced both in extension and in thickness and shows a variety of lamination defects. The archicortex of these embryos is also heavily affected: the dentate gyrus is missing, the hippocampus and the medial limbic cortex are greatly reduced in size. In addition, the olfactory bulb is disorganized and the olfactory nerve fails to project to it (Pellegrini et al 1996, Yoshida et al 1997).

A suggestion for an involvement of *Emx2* in the organization of the neocortical neuronal migration came from the finding that mutations of this gene in humans are responsible for some cases of schizencephaly (Brunelli et al 1996, Faiella et al 1997, Granata et al 1997). Schizencephaly is a rare brain developmental malformation due to neuronal migration disorder and it is characterized by full thickness clefts of the cortical layer of cerebral hemispheres, allowing communication between the ventricle and pericerebral subarachnoid space.

Whereas early analysis detected *Emx2* expression in proliferating cortical neuroblasts, subsequent studies of our group (Mallamaci et al 1998) also detected the presence of EMX2 protein in Cajal-Retzius cells in the cortical marginal zone of late gestation mouse embryos. Cajal-Retzius cells are a transient cell populations playing a major role in orchestrating the radial migration of cortical neurons (Marín-Padilla 1998, del Río et al 1995), partly through the protein product of the *reelin* gene (D'Arcangelo et al 1995, 1997, D'Arcangelo & Curran 1998, Ogawa et al 1995).

As it is conceivable to hypothesize a role of the EMX2 protein in establishing and/or maintaining the identity of Cajal-Retzius cells and in the exploitation of their function, we decided to investigate in depth the cortical cytoarchitecture of *Emx2* null mutant mouse embryos (Mallamaci et al 2000). The formation of the cerebral cortex is a complex process characterized by many steps (Bayer & Altmann 1991) (Fig. 1). Early postmitotic neurons accumulate at the marginal edge of the cortical wall, forming the so-called primordial plexiform layer. Then, later-born neurons climb along fascicles of radial glia and infiltrate the primordial plexiform layer. They split it into the more superficial marginal zone and the deeper subplate (Ghosh 1995) and accumulate between them, making up the cortical plate (Marín-Padilla 1998).

Mid-gestation *Emx2* null embryos lack Cajal-Retzius cells (Mallamaci et al 2000). It is conceivable that, in the absence of *Emx2* products, these cells fail to be born, to survive or to undergo the appropriate differentiation programme. The severe impairment of the neuronal radial migration that is observed might in turn be a consequence of this phenomenon. In addition to that, we found that specific neuronal subpopulations of the subplate are selectively affected in *Emx2* null embryos at axial locations just corresponding to those cortical areas, namely posterior and medial, which normally express *Emx2* at the highest levels (Mallamaci et al 2000).

The absence of Cajal-Retzius cells in mid-gestation mutant embryos was demonstrated by monitoring the expression of four molecules: calretinin, GAP43, reelin and EMX1. In the mutant embryos, calretinin, GAP43 and reelin

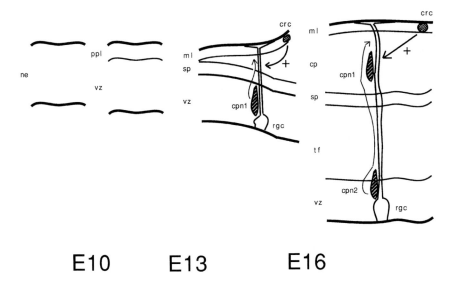

E10   E13   E16

FIG. 1.   Scheme of cortical formation and lamination. At the beginning only a germinative neuroepithelium (ne) is present. A few days later a primordial plexiform layer (ppl) is formed which is subsequently split in an external marginal layer (ml) and a subplate (sp) by the ingression of outward migrating cortical plate neurons (cpn) deriving from the internal ventricular zone (vz). These neurons migrate along the fascicles of fibres of the radial glial cells (rgc) and respond to signals (+) emanating, among other things, from the Cajal-Retzius cells (crc) located in the marginal layer. Neurons arriving later (cpn2) into the cortical plate (cp) through the transitional field (tf) migrate past those (cpn1) arrived in an earlier phase. Approximate stages of mouse development are indicated. In *reeler* mutants, cortical plate neurons are not able to penetrate the preplate, which does not get split and gives rise to the so-called superplate. In *Emx2* null mutant embryos these cells partially retain this ability, so that a subplate can be distinguished from the subpial layer, but fail to laminate properly. E, embryonic day.

mRNA signals were absent in the entire neocortical marginal zone. Down-regulation of these markers could be consequence of defective birth/survival of cells that normally express them or that are targeted by axons expressing one of them. Alternatively, these cells, even if born and surviving, could have not followed their proper differentiation programmes. In order to distinguish between these two alternatives, we administered BrdU to pregnant mothers at E12.0, a time at which both subplate cells and late Cajal-Retzius cells are co-generated (Bayer & Altmann 1991, Meyer & Fairèn 1996), and scored the radial distribution of BrdU immunoreactive cells in the embryonic neocortex around the end of gestation. In wild-type embryos about one-half of pulse-labelled cells were aligned underneath the pia, while the rest were loosely clustered at an intermediate radial level, roughly corresponding to the subplate. In mutant embryos, no BrdU positive cells were detectable at the marginal edge of the cortical wall. All of them lay around the putative subplate level. The absence of Cajal-Retzius cells in the marginal zone suggests that radial migration of late-born neurons should be deeply perturbed in these mice, possibly in a *reeler*-like way (Caviness et al 1988).

In order to systematically compare the migratory behaviour of cells fated to form the cortical plate in *Emx2* null and wild-type embryos, we pulse-labelled neurons born at E12.0, E13.5 and E15.0 by BrdU and scored their radial distribution throughout the neocortical wall at about the end of gestation. The analysis was performed on E19 null mutant, wild-type and *reeler* embryos and the radial distribution of cells born at different times in embryos of different genotypes was analysed. The radial distribution of normal neurons born at E12.0 gave rise to two peaks, a sharp marginal one, corresponding to Cajal-Retzius cells, and a smooth intermediate one, at the level of the subplate. Both E13.5 and E15.0 curves displayed one peak, falling in between the two E12.0 peaks; in addition, the E15.0 peak was superficial to the E13.5.

In mutant embryos, the E12.0 graph gave rise only to a smooth peak, lying at the presumptive subplate level; no evidence of the normal marginal peak was detected, suggesting that no Cajal-Retzius cells were under the pia in these embryos. In the same embryos, both the E13.5 and E15.0 curves were flatter than expected. A fraction of E13.5 and E15.0 born cells were located marginally to the smooth intermediate E12.0 peak, which suggests that cortical plate neurons were still able to infiltrate the primordial plexiform layer. However, both E13.5 and E15.0 main peaks were displaced toward the ventricular side. Moreover, the E13.5 and E15.0 populations were intermingled and the inside-outside distribution was hardly detectable.

This picture is reminiscent of the *reeler* mutant cerebral cortex (Caviness et al 1988). However, in *reeler* mutants cortical plate neurons are not even able to penetrate the preplate, which does not split, and gives rise to the so-called superplate. In *Emx2* null embryos cortical plate neurons partially retain this ability, so that a subplate can be distinguished from the subpial layer. Cajal-Retzius cells are the only effectors of the reelin function in the developing cerebral cortex around E15. Hence, the *reeler*-like phenotype exhibited by *Emx2* null mutant embryos seems to be a consequence of the absence of these cells in the mutant animals at this stage. However, despite the absence of Cajal-Retzius cells at E15.5, the *Emx2* mutant migratory phenotype is less severe than the *reeler* one, suggesting that the reelin neocortical function in the *Emx2* mutant mice is not completely impaired.

We found that at E11.5 the distribution and number of reelin-expressing cells in the cortical wall of *Emx2* mutant and wild-type embryos were similar, especially in the neocortical anlage. Among Cajal-Retzius cells, primary and secondary populations have been distinguished in the rat. Primary Cajal-Retzius cells appear in the primitive neuroepithelium at stages corresponding to mouse E9.5–E11.0, and disappear completely about three days later. Secondary Cajal-Retzius cells are born at stages corresponding to mouse E11.0–E12.5 and disappear at different moments during pre- and postnatal life (Meyer & Fairèn 1996). We believe that the Cajal-Retzius cells absent in *Emx2* null mutant mice at E15.0 represent secondary Cajal-Retzius cells. Conversely, the Cajal-Retzius cells present in mutant embryos at E11.5 should be primary Cajal-Retzius cells. Alternatively, they may represent newborn secondary Cajal-Retzius cells fated to die prematurely.

In *reeler* mutants, specific morphological abnormalities in radial glial cells and cortical plate neurons are invariably associated with migration defects (Caviness et al 1988). Similar anomalies can be detected in *Emx2* null mutants. During their translocation to the cortical plate, radially elongated neurons climb along the surface of bundles of radial glia (Fig. 1). In wild-type animals, at E16.5, the average diameter of glial bundles varies dramatically at the level of the subplate as a consequence of the defasciculation process which normally occurs to radial glia between E15 and E17 in the mouse. The defasciculation process consists of the separation of the three to 10 single cell fibres forming each primary glial fascicle and is associated with the penetration of each primary fascicle by radially migrating neurons (Caviness et al 1988). In *Emx2* null mutant embryos the glial defasciculation is also defective. The glial fascicles display a more graded ventricular to marginal reduction in size and it is possible to find plenty of thick bundles reaching the marginal zone. It is conceivable that the absence of reelin protein is the common cause of this phenomenon in both *reeler* and *Emx2* mutants.

Migrating cortical neurons of *Emx2* null mutant embryos also showed abnormalities, in shape and packaging pattern. At E16, the neocortical plate of mutant mice contained prevalently rounded neurons, instead of the characteristic fusiform cells populating the wild-type plate. In addition, these neurons were loosely clustered and the mutant cortical plate lacked the tight and palisade-like architecture characterizing the normal plate.

In summary, in *Emx2* null mutant mice, late phases of neocortical lamination are selectively impaired. In these mutant embryos, the primordial plexiform layer is formed and early neocortical Cajal-Retzius cells are normally born. Cortical plate neurons infiltrate the primordial plexiform layer and split it into the marginal layer and the subplate. Subsequently, however, at a time when primary wild-type Cajal-Retzius cells have entirely been replaced by secondary cells, the mutant neocortical marginal zone appears to be completely depleted of any Cajal-Retzius cell. Because of the consequent lack of reelin signalling, late development of radial glia is perturbed and late-born cortical plate neurons fail to overcome earlier-born ones, so that cortical layers are not laid down in an appropriate way. It could be hypothesized that similar phenomena could also underlie specific neocortical defects observed in human schizencephalic patients carrying mutations in the *Emx2* locus and other congenital migration defects.

Acknowledgements

The experimental part of this work was supported by grants from the European Community Biotechnology and Biomedicine Programmes, the Telethon-Italia Programme, the Italian Association for Cancer Research and the Arinenise-Harvard Foundation.

References

Bayer SA, Altmann J 1991 Neocortical development. Raven Press, New York

Boncinelli E 1994 Early CNS development: *distal-less* related genes and forebrain development. Curr Opin Neurobiol 4:29–36

Boncinelli E 1999 *Otx* and *Emx* homeobox genes in brain development. Neuroscientist 5: 64–172

Briata P, Di Blas E, Gulisano M et al 1996 EMX1 homeoprotein is expressed in cell nuclei of the developing cerebral cortex and in the axons of the olfactory sensory neurons. Mech Dev 57:169–180

Brunelli S, Faiella A, Capra V et al 1996 Germline mutations in the homeobox gene *EMX2* in patients with severe schizencephaly. Nat Genet 12:94–96

Caviness VS Jr, Crandall JE, Edwards MA 1988 The *reeler* malformation: implications for neocortical histogenesis. In: Jones EG, Peters A (eds) Cerebral cortex, vol 7: Development and maturation of the cerebral cortex. Plenum Press, New York, p 59–89

D'Arcangelo G, Curran T 1998 *Reeler*: new tales on an old mutant mouse. Bioessays 20:235–244

D'Arcangelo G, Miao GG, Chen SC, Soares HD, Morgan JI, Curran T 1995 A protein related to extracellular matrix proteins deleted in the mouse mutant *reeler*. Nature 374:719–723

D'Arcangelo G, Nakajima K, Miyata T, Ogawa M, Mikoshiba K, Curran T 1997 Reelin is a secreted glycoprotein recognized by the CR-50 monoclonal antibody. J Neurosci 17:23–31

del Río J, Martinez A, Fonseca M, Auladell C, Soriano E 1995 Glutamate-like immunoreactivity and fate of Cajal-Retzius cells in the murine cortex as identified with calretinin antibody. Cereb Cortex 5:13–21

Faiella A, Brunelli S, Granata T et al 1997 A number of schizencephaly patients including two brothers are heterozygous for germline mutations in the homeobox gene *EMX2*. Eur J Hum Genet 5:186–190

Fernández AS, Pieau C, Repérant J, Boncinelli E, Wassef M 1998 Expression of the *Emx-1* and *Dlx-1* homeobox genes define three molecularly distinct domains in the telencephalon of mouse, chick, turtle and frog embryos: implications for the evolution of telencephalic subdivisions in amniotes. Development 125:2099–2111

Ghosh A 1995 Subplate neurons and the pattering of thalamo-cortical connections. In: Development of cerebral cortex. Wiley, Chichester (Ciba Found Symp 193) p 150–165

Granata T, Farina L, Faiella A et al 1997 Familial schizencephaly associated with *EMX2* mutation. Neurology 48:1403–1406

Gulisano M, Broccoli V, Pardini C, Boncinelli E 1996 *Emx1* and *Emx2* show different patterns of expression during proliferation and differentiation of the developing cerebral cortex. Eur J Neurosci 8:1037–1050

Mallamaci A, Iannone R, Briata P et al 1998 EMX2 protein in the developing brain and olfactory area. Mech Dev 77:165–172

Mallamaci A, Mercurio S, Muzio L et al 2000 The lack of *Emx2* causes impairment of Reelin signalling and defects of neuronal migration in the developing cerebral cortex. J Neurosci 20:1109–1118

Marín-Padilla M 1998 Cajal-Retzius cells and the development of the neocortex. Trends Neurosci 21:64–71

Meyer G, Fairèn A 1996 Different origins and developmental histories of transient neurons in the marginal zone of the fetal rat cortex. Soc Neurosci Abstr 22:1014

Morita T, Nitta H, Kiyama Y, Mori H, Mishina M 1995 Differential expression of two zebrafish *emx* homeoprotein mRNAs in the developing brain. Neurosci Lett 198:131–134

Northcutt RG, Kaas J 1995 The emergence and evolution of mammalian neocortex. Trends Neurosci 18:373–379

Ogawa M, Miyata T, Nakajima K et al 1995 The *reeler* gene-associated antigen on Cajal-Retzius neurons is a crucial molecule for laminar organization of cortical neurons. Neuron 14:899–912

Pannese M, Lupo G, Kablar B, Boncinelli E, Barsacchi G, Vignali R 1998 The *Xenopus Emx* genes identify presumptive dorsal telencephalon and are induced by head organizer signals. Mech Dev 73:73–83

Patarnello T, Bargelloni L, Boncinelli E, Spada F, Broccoli V 1997 Evolution of *Emx* genes and brain development in vertebrates. Proc R Soc Lond B Biol Sci 264:1763–1766

Pellegrini M, Mansouri A, Simeone A, Boncinelli E, Gruss P 1996 Dentate gyrus formation requires *Emx2*. Development 122:3893–3898

Shimamura K, Martinez S, Puelles L, Rubenstein JLR 1997 Patterns of gene expression in the neural plate and neural tube subdivide the embryonic forebrain into transverse and longitudinal domains. Dev Neurosci 19:88–96

Simeone A, Gulisano M, Acampora D, Stornaiuolo A, Rambaldi M, Boncinelli E 1992a Two vertebrate genes related to *Drosophila empty spiracles* gene are expressed in embryonic cerebral cortex. EMBO J 11:2541–2550

Simeone A, Acampora D, Gulisano M, Stornaiuolo A, Boncinelli E 1992b Nested expression domains of four homeobox genes in the developing rostral brain. Nature 358:687–690

Simeone A, Acampora D, Mallamaci A et al 1993 A vertebrate gene related to *orthodenticle* contains a homeodomain of the *bicoid* class and demarcates anterior neuroectoderin in the gastrulating mouse embryo. EMBO J 12:2735–2747

Striedter GF 1997 The telencephalon of tetrapods in evolution. Brain Behav Evol 49:179–213

Yoshida M, Suda Y, Matsuo I et al 1997 *Emx1* and *Emx2* functions in the development of dorsal telencephalon. Development 124:101–111

DISCUSSION

Parnavelas: How do the thickness and the lamination of the cortex in these knockout mice compare with the wild-type?

Boncinelli: We haven't yet looked at this because we have been focusing on the Cajal-Retzius cells, but we will.

Parnavelas: Is *Emx2* expressed in the ganglionic eminence of wild-type mice?

Boncinelli: Yes, it is expressed in a region of the ganglionic eminence.

Goffinet: You mentioned that Cajal-Retzius cells were present at embryonic day (E) 11.5, but had disappeared by E15.5. Did you show this by *in situ* hybridization experiments?

Boncinelli: Yes, we used a variety of probes including calretinin and reelin mRNA, but also antibodies against GAP43, MAP2, TuJ1 and EMX1.

Goffinet: Could this be explained by the premature death of Cajal-Retzius cells?

Boncinelli: It is possible, but we do not know for sure. There are at least two waves of migration. The first starts at around E9–10. These early cells appear to migrate to the correct place in *Emx2* null mice. In contrast, those born at E12 are not detectable in the developing brain of these mutant mice, possibly because they fail to migrate properly.

Molnár: Meyer et al (1998) have evidence that there are at least two populations of marginal zone cells, which is in agreement with your data.

Boncinelli: This may explain why *Emx2* null mutant mice have a severely deformed hippocampus and completely lack the dentate gyrus. It is known that Cajal-Retzius cells, especially a late population of these cells, play a pivotal role in directing the development of these structures. Lack of Cajal-Retzius cells might destroy their morphology.

Goffinet: The knockout animals seemed to have an almost normal molecular layer. Did you check the maturation gradient in the cortical plate of those animals? Is it inside-out or outside-in?

Boncinelli: We did many labelling studies, and we found that in our *Emx2* null mice these neurons properly migrate past the putative subplate and only subsequently their migration pattern is disorganized. Conversely, in *reeler* mice, migrating cortical neurons entirely fail to migrate past the subplate. By definition, *reeler* mice do not have any reelin production. In our case there is some reelin expressed and present in (almost) appropriate locations. They appear to suffer for both a shortage of reelin and other migration guidance clues.

Levitt: You showed a slide of homologous areas that were based on your gene expression pattern data. The dark blue area almost seemed non-existent, which I find confusing because it suggests that this region is an evolutionary new structure.

Karten: We just don't yet have the gene that marks it.

Puelles: You get portions of the claustrum and portions of the amygdala developing in this position.

Karten: But Pat Levitt is asking about the gene, and not the interpretation.

Puelles: But the fact that the region we are discussing is small and has momentarily only been distinguished by the lack of expression of *Emx1*, as compared to the rest of the pallium, doesn't mean that there are no identifiable neurons in it. The claustrum can be traced in mammals all the way from the orbital cortex down to the posterior, amygdaloid part of the telencephalon, so it may be rather thin lateromedially, but is actually rather long. It thus builds a sizeable mass, to which lateral parts of the amygdala contribute sizeable additions, all within the same molecularly defined compartment.

Levitt: This suggests that the basal medial nucleus in the amygdala, claustrum and globus pallidus rostrally all originate in the medial ganglionic eminence. However, the projection patterns are completely different. for example, the projections from the prefrontal cortex to the claustrum are enormous in rodents, and there are other direct projections into the basal medial nucleus.

Puelles: Corticoclaustral interconnections seem to predominate in the dorsolateral part of claustrum that falls inside the *Emx1*-expressing domain, which is perhaps comparable as a field to the avian hyperstriatum ventrale. There are also projections interconnecting the hyperstriatum ventrale and the overlying cortical fields in birds. On the other hand, ventromedial parts of claustrum and associated amygdala counterparts may be more significantly projected upon by dorsal thalamic neurons, as happens in the avian and reptilian neostriatum; these are the areas negative for *Emx1*. I'm not saying that the picture is completely clear, but that we should seriously consider the predictions generated if this molecularly distinct region present in all tetrapods, as stressed here by Boncinelli, is homologous to the claustrum and amygdala in mammals.

Boncinelli: It is clear that we don't inherit biological structures. Our hand does not derive from the hand of an ancestor. It is built and shaped by some genes under the control of other regulatory genes. Only these genes and genetic circuits derive from those of our ancestors. But it is also true that we certainly do not inherit single genes. We inherit genomes that are in turn, presumably, a collection of genetic circuits and networks. In this light, it is appropriate to study gene networks, not single genes. One gene may fail to provide useful and reliable evolutionary information, as may do two or three genes. On the contrary, the study of genetic networks should not be misleading.

Puelles: If we find out that many of these gene patterns are comparable within a consistent topology (*Bauplan*), then we will tend to accept homology, even if it is only field homology. If we still find differential projections, then this suggests that evolution of a differential aspect of structure and function may have occurred; projections can be added or lost more easily than viable constellations of multiple gene expression patterns. It is possible to have a different projection pattern within homologous structures, since homology does not require identity of all aspects. It is not necessary to define everything from the projections. We need to start from the bottom and look at early specification and subsequent cell differentiation and migration patterns, then axonal navigation. Finally we must consider the minor aspects of cell type identity, which may be conserved or not conserved (this also depends on our viewpoint, techniques and interests, since these will drive us to attend to cellular characters that confirm our assumptions and may lead us to disregard the essential aspects selected for in evolution).

Hunt: If I have two cells, one in the bird and one in the mammal, that express 20 identical genes during development, what can I predict about the relationship between these two cells?

Papalopulu: If you observe that one cell in two different organisms expresses 20 identical genes, you can probably not predict that much. However, if there is a pair of cells in two organisms that express the same genes between organisms, one way to decide if they are homologous is to find out whether these two cells have the same topographical relationship to each other during development. For example, *distalless*, *Pax6* and *Emx1* are expressed in different organisms, and some areas in which these genes are expressed are larger in some organisms than in others. However, it is likely that the sizes and relative locations of these territories will be the same if you look further back in development, which in my mind would strengthen the argument that these genes mark homologous structures.

Rakic: The problem with this is that the correct experiment is to look at what has happened in the common ancestor, and this is not possible to do.

Puelles: I agree with Nancy Papalopulu that the expression patterns of the 20 identical genes doesn't tell you everything, because it is also necessary to consider the location of the cells at the moment their fate was specified (obviously, it is not always easy to know this).

Karten: When you say 'location' you are also acknowledging that migration is a major player. However, if you leave out the absolute location, what conclusions can you draw? How many factors have to be coincident, and if they're not coincident what does this mean?

Puelles: Final, postmigratory position is one thing and primary postmitotic position is another. Consideration of position simply cannot be left out for any meaningful conclusions. My impression is that conserved aspects of neuronal populations reside more in their primary than in their secondary positions (i.e. γ-aminobutyric [GABA]ergic nature of cortical interneurons stemming from the subpallium), although I am well prepared to accept less clear-cut situations (i.e. specification occurring midways along a migratory route, or only after stabilization). But throughout evolution you would expect to see both conserved and differential aspects in different species.

Karten: Is a developmental character, of the sort we are talking today, necessarily better than, 20 phenotypic characters, for example, or the presence or absence of a specific gene expression pattern?

Puelles: In such a multivariable and multiparametric system as is the developing brain, this question seems to me largely undecidable. It is clear that any single gene or single phenotypic property (which can be theoretically reduced to various gene functions) can appear repeatedly in different positions throughout the brain; this leads us to analogy, not to homology. For hypothesizing homology, we are compelled to identify and analyse as many variables as possible, relate them to positions and try to develop the simplest explanatory theories. I regard (positionally and molecularly) homologous developmental histogenetic fields in the neural tube as parameters of brain development (they can change, but do so slowly, or rarely at all in evolution). Cell populations and their connective or pharmacological phenotypes would represent variables regulated more dynamically, according to the specific parameters in their primary (possibly also in secondary) environments.

Wolpert: If you have 10 000 identical active genes and one different active gene in two cells, are those two cells the same?

Karten: I don't know. That's the issue we need to discuss. What are the factors that constitute evolutionary change and speciation?

Reiner: A possible example of this is the malleus and incus. They will have a common expression pattern for some genes, because they have a common ancestry, but they will also have different patterns of expression. The key issue is that we use these similarity traits to try to make inferences about homology, but if we really want to talk about homology, ideally we should try to track traits back to the common ancestor. Sometimes this is possible, and sometimes it is not.

Rakic: In answer to the question concerning the factors that constitute evolutionary change and cortical speciation, you are really talking about the comparison between phenotype and genes, and I would say that they are two different entities. The study of the phenotype is a consequence, whereas the study of the gene expression is the study of the consequences of change.

Karten: No. As histologists we are talking about a marker that we stain for. One of the points that stands out in Eduardo Boncinelli's paper is that the *Emx1* staining pattern in the cortex is splotchy, and there are layers, for example layer I and IV, that are completely unstained.

Parnavelas: This is because these layers contain predominantly interneurons, and *Emx1* is expressed by pyramidal neurons.

Karten: Layer IV contains mostly thalamic recipient neurons, and at a cell-by-cell level the question is, does a thalamic recipient neuron express *Emx1*? When we are trying to relate things that are going on at the level of a single cell, we have to consider the limitations of the *in situ* analysis, i.e. can we draw any conclusions about a single cell if it does not express *Emx1*?

Puelles: May I point out the danger of reasoning on gene expression as a static feature? What we need is the complete life history of any interesting cell population in terms of up-regulated and down-regulated genes and the corresponding consequences. Any cell may not express one gene at a given moment, while doing so at earlier or later periods; but that cell will always express other genes and the changes in the sequence should be amenable to scientific analysis. If you compare the expression of *Emx1* and *Emx2* at the lateral ventricle of the telencephalon, you see that there are ventricular cells near the lateral angle of the ventricle, just outside the subpallium, that express *Emx2*, but not *Emx1*. This difference identifies for us a particular territory that has a different gene expression pattern from the surrounding area throughout subsequent stages of development. Even though the absence of *Emx1* itself may not be causally important for local cell fate specification and histogenesis, this result distinctly identifies an area where development is proceeding differentially, undoubtedly due to the undiscovered agency of other genes. Therefore, neurons generated here reasonably can be expected to represent a distinct cell group both in mammals and in other vertebrates. This is the place where we would expect the origin of neural migrations postulated by Harvey Karten to invade layer IV in some mammalian cortical areas, as long as the argument starts with the consideration of avian and reptilian cell derivatives of this molecularly and topologically characteristic area to be possible homologues of these mammalian layers. Harvey's emphasis on the presence of *Emx1*-negative cells in layer IV of some cortical areas shows how his argument may still be held to be consistent with our novel state of molecular understanding of this particular telencephalic primordium (other possibilities: the layer IV cells may not express any pallial markers, and they perhaps arose in the subpallium — then their connections are all wrong; or the cells did express *Emx1* early on and secondarily down-regulated its signal — their connections at best would be analogous to those of consistently *Emx1*-negative neurons). It seems that the critical question to be posed experimentally is whether any cells arising in the mammalian *Emx1*-negative pallial domain migrate into the predicted cortical layers. Such evidence is not yet available.

Reference

Meyer G, Soria JM, Martínez-Galán JR, Martín-Clemente B, Farién A 1998 Different origins and developmental histories of transient neurons in the marginal zone of the fetal and neonatal rat cortex. J Comp Neurol 397:493–518

Intrinsic and extrinsic control of cortical development

John L. R. Rubenstein

Nina Ireland Laboratory of Developmental Neurobiology, Center for Neurobiology and Psychiatry, Department of Psychiatry and Programs in Neuroscience, Developmental Biology and Biomedical Sciences, 401 Parnassus Avenue, University of California at San Francisco, CA 94143-0984, USA

Abstract. Recent advances in the study of cerebral cortical early development are described in this chapter. The role of the anterior neural ridge in regulating telencephalon induction in the neural plate is discussed, followed by a review of the evidence for the roles of ventral, rostral and dorsal patterning centres in regulating regionalization of the telencephalon. The patterning centres produce secreted molecules (SHH, FGF, BMP, WNT) that regulate the expression of transcription factors which control regional identity, cell type specification, proliferation and differentiation. These intrinsic patterning mechanisms appear to be sufficient to generate much of the regional organization of the cerebral cortex present in newborn mice. While intrinsic mechanisms have a major role in cortical regionalization and in the production of cortical projection neurons, many cortical interneurons are derived from the basal ganglia and then migrate into the cerebral cortex. Furthermore, thalamic afferents appear to have an important role in maturation of the postnatal rodent cortex. Thus, both intrinsic and extrinsic mechanisms control development of the cerebral cortex.

2000 Evolutionary developmental biology of the cerebral cortex. Wiley, Chichester (Novartis Foundation Symposium 228) p 67–82

The vertebrate cerebral cortex, or pallium, is organized into several large functional subdivisions: the medial, dorsal, lateral and ventral pallium (Puelles et al 2000), each with distinct histologies and connectivities. In mammals, the medial pallium, or archicortex, includes the hippocampal region, with its characteristic pyramidal cell layer in the CA fields and the granule cells of the dentate gyrus. The mammalian dorsal pallium corresponds to the neocortex, which has six principal layers. The lateral pallium corresponds to the paleocortex (the primary olfactory cortex), which has a single layer of pyramidal neurons. It is postulated that the ventral pallium includes the claustrum (Puelles et al 2000), which is a deep grey matter structure. Further subdivisions are found in each of these regions, as exemplified by the prefrontal, motor and sensory regions within the neocortex. During evolution, the cortex has shown a disproportionate increased surface area with respect to most other brain regions. For instance, in rodents and small-brained insectivores, the cortical surface area is 3–5 cm^2/hemisphere, whereas in humans it is roughly 1100 cm^2/hemisphere (reviewed in Northcutt & Kaas 1995). With this expansion there has been a similar increase in the computational abilities of the brain. Given the central role of the cerebral cortex in normal and abnormal cognition, it is important to understand how this structure forms, as changes in its developmental programmes may underlie variance of cognitive functions within species, neuropathological states or its evolution. In this chapter I will briefly survey some recent studies that are beginning to elucidate the sequence of processes that assemble the cerebral cortex.

Induction of the telencephalon

The cerebral cortex encompasses most of the dorsomedial aspect of the telencephalon. The telencephalon is a vesicular outgrowth from the dorsolateral walls of the prosencephalon. Fate mapping studies indicate that the anlage of the telencephalon lies at the rostrolateral aspect of the neural plate (reviewed in Rubenstein et al 1998). The anlage of the cerebral cortex maps to the caudolateral parts of the telencephalic primordium (Rubenstein et al 1998, Fernández et al 1998). There is evidence that the rostral edge of the neural plate (the anterior neural ridge [ANR]) is the location of a patterning centre that regulates the growth and regionalization of the forebrain. Production of fibroblast growth factor (FGF)8 in the ANR is implicated in regulating expression of BF1 (Shimamura & Rubenstein 1997, Ye et al 1998), a transcription factor required for regionalization and proliferation of the telencephalon (Xuan et al 1995). Whether telencephalic subdivisions are beginning to be specified within the neural plate is unknown, although there is regionalized expression of regulatory genes within this tissue (see Shimamura et al 1995, Shimamura & Rubenstein 1997, Rubenstein et al 1998). Following induction of markers of the telencephalon in the neural plate (e.g. BF1), neurulation leads to the formation of the neural tube and, subsequently, evagination of the telencephalic vesicles.

Patterning of the telencephalon

The telencephalon can be conceived as having two major subdivisions: the pallium (or cortex) and the subpallium (or basal telencephalon) (Puelles et al 2000). The dorsal midline and paramedian tissues, which are continuous with the roof plate of more posterior regions of the CNS (Rubenstein et al 1998), give rise to the commissural plate and the choroid plexus. Adjacent to the choroid plexus is the anlage of the fimbria. Flanking the fimbria is the anlage of the dentate gyrus and

CA field of the hippocampus; we define this region as the medial pallium (Puelles et al 2000). According to our model, the dorsal, lateral and ventral pallium are distinct cortical zones. The ventral pallium abuts the pallial/subpallial boundary. The subpallium consists of three subdivisions. From dorsal to ventral these are: striatum, pallidum and anterior entopeduncular areas (AEP). Rostrally, many of the telencephalic zones converge into the septal area, whereas caudally, they converge into the amygdalar complex (Puelles et al 2000).

Regionalization of the embryonic telencephalon is regulated by the production of morphogens from patterning centres (Rubenstein et al 1998, Rubenstein & Beachy 1998). Ventral specification of the telencephalon is regulated by sonic hedgehog (SHH). There are probably at least two phases of SHH function: early ventral specification is controlled by the axial mesendoderm (prechordal plate, which underlies the neural plate); later ventral specification may be controlled by SHH expression in rostral regions of the telencephalic stalk (preoptic area and AEP). SHH mutants, which lack the function of both patterning centres, also lack basal telencephalic structures (Chiang et al 1996); the organization of the rudimentary cortex that forms in these holoprosencephlic mice has not been scrutinized. Gain-of-function experiments show that SHH can ventralize explants of the cerebral cortex, and thus has the potential to pattern the pallium (Kohtz et al 1998). Analysis of mice with a mutation in the *Nkx2.1* homeobox gene (Sussel et al 1999) shows that telencephalic expression of SHH is almost eliminated; despite this, histogenesis of the major cortical subdivisions appears normal. Thus, perhaps SHH expression within the telencephalon is not essential for cortical regionalization.

We have proposed that there is a telencephalic patterning centre at its rostral midline (Shimamura & Rubenstein 1997). In the neural plate, this region is called the ANR, which in the neural tube becomes several tissues including the septum. This tissue expresses FGF8, and may regulate anteroposterior patterning and growth of the telencephalon (P. Crossely, E. Storm, G. Martin & J. L. R. Rubenstein, unpublished results 1999).

The dorsal midline and paramedian tissues of the telencephalon are another source of patterning signals that regulate regionalization of the dorsal telencephalon. This patterning centre expresses bone morphogenetic proteins (BMPs), growth differentiation factors (GDFs) and WNTs (Parr et al 1993, Furuta et al 1997, Tole et al 1997, Tole & Grove 1999), molecules that regulate dorsal patterning of caudal CNS tissues (Ikeya et al 1997, Lee et al 1998). Gain-of-function experiments also show that BMPs can dorsalize the chick telencephalon (Golden et al 1999, Y. Ohkubo & J. L. R. Rubenstein, unpublished results 1999). A role for WNT proteins is suggested by hippocampal defects found in mice lacking the LEF1 transcription factor, a protein implicated in WNT signal transduction (Galceran et al 2000).

Additional patterning centres flanking, and within, the telencephalon may control regionalization of the cerebral cortex. There is evidence that the olfactory placode and its associated mesenchyme are required for olfactory bulb development through retinoid-mediated mechanisms (Anchan et al 1997). It is conceivable that regionalization is also regulated by boundary regions within the telencephalon, such as the cortical/subcortical boundary.

Signals from the patterning centres are transduced into information controlling regional fate by inducing or repressing expression of transcription factors. SHH-mediated patterning of the ventral telencephalon is controlled via the induction of *Nkx2.1* (Sussel et al 1999); mutation of *Nkx2.1* dorsalizes pallidal parts of the basal ganglia (Sussel et al 1999).

Signals from dorsal patterning centres are believed to regulate the expression of *Gli3*, *Emx1*, *Emx2* and *Lef1*, which encode transcription factors. Mutations of these genes affect development of dorsal telencephalic structures, such as parts of the hippocampus, corpus callosum and choroid plexus (Qiu et al 1996, Pelligrini et al 1996, Yoshida et al 1997, Grove et al 1998, Tole & Grove 1999). Furthermore, BMPs can repress BF1; this repression may explain why BF1 transcripts are not detected from dorsomedial regions of the telencephalon (Furuta et al 1997). In addition, BF1 may be able to repress the expression of BMP4; in BF1 mutants, BMP4 is expressed throughout the telencephalic vesicle (Dou et al 1999).

Anteroposterior patterning may be transduced in part via the *Otx1* and *Otx2* homeobox genes, which are expressed in the midbrain and forebrain. Mice that are $Otx1^{-/-}$ and $Otx2^{+/-}$ exhibit posteriorization of the forebrain (Acampora et al 1997). This may occur because of a rostral shift in the position of the midbrain/hindbrain patterning centre (the isthmic organizer) and/or due to an intrinsic role of the *Otx* genes in regulating forebrain regional specification. In addition, there is evidence that one of the functions of FGF8 in the isthmic organizer is to regulate *Otx* expression (Martinez et al 1999). Thus, we are investigating whether FGF8 expression in the ANR (and commissural plate) may regulate forebrain development through controlling expression of the *Otx* genes.

Evidence that prenatal regionalization of the rodent cortex does not require thalamic input

While telencephalic patterning centres can regulate telencephalon regionalization, there is little information demonstrating their roles in controlling the organization of the cortex. On the other hand, there is evidence that regionalization of the cortex is controlled by intrinsic patterning mechanisms based on transplantation and explant culture methods. These experiments indicate that expression of regional molecular markers of the neocortex (Cohen-Tannoudji et al 1994, Nothias et al

1998), archicortex (Tole & Grove 1999) and lateral limbic cortex (Levitt et al 1997) do not depend upon extrinsic factors, such as afferent axons.

Other investigators have studied cortical development in mice lacking thalamic afferents. For instance, Wise & Jones (1978) performed a thalamotomy in newborn rats and found that a histological characteristic of layer IV (dense granule cell aggregates) in the somatosensory cortex was maintained. More recently, we have studied neocortical regionalization in *Gbx2* mutant mice that lack thalamocortical fibres (Miyashita-Lin et al 1999). The *Gbx2* homeobox gene is required for thalamic differentiation and the elaboration of the thalamocortical tract (Miyashita-Lin et al 1999). We analysed the organization of the cortex in newborn *Gbx2* mutant mice using a panel of gene markers, whose expression defines cortical subdivisions (Miyashita-Lin et al 1999, Rubenstein et al 1999). We found that the expression of these genes was not affected by the absence of thalamic afferents. This supports the proto-map model of Rakic (1988) which postulates that regionalization of the cortex is regulated by intrinsic factors.

While thalamic afferents may not be essential for generating a coarse map of cortical subdivisions, they probably have a major role in maturation of the neocortex. For instance, heterotopically transplanted immature cortical tissue can develop efferent projections and histology that characterize the local cortical region (e.g. Schlagger & O'Leary 1991). In addition, altering the anatomy (ablation of whisker pads or ventrobasal thalamus) or the function of the thalamic inputs to the rodent somatosensory (e.g. pharmacological disruption or mutations that affect monoamine oxidase, adenylate cyclase, serotonin, NMDA receptor), disrupt histogenesis of the somatosensory cortex (Killackey et al 1995, Cases et al 1996, Welker et al 1996, Iwasato et al 1997, Abel-Majid et al 1998). In addition, thalamic inputs are implicated in regulating the expression of the H-2Z1 transgene in the somatosensory cortex (Gitton et al 1999). These results probably reflect the observation that the functional organization of the neocortex in young mammals is plastic (Hubel 1988). In addition, they suggest that anatomical and/or functional changes in axonal inputs to the neocortex can play a role in modifying existing, and generating new, neocortical subdivisions (Krubitzer 1995, Innocenti 1995).

Cortical and subcortical origins of neurons of the cerebral cortex

While cortical regionalization may be regulated by intrinsic factors, its cellular composition is not entirely derived from the cortical ventricular zone. Thus, there is evidence that cortical projection neurons migrate radially from the cortical ventricular zone to the cortical mantle (Tan et al 1998), whereas cortical γ-aminobutyric acid (GABA)-ergic interneurons are derived from the ventricular zones of the basal telencephalon and arrive in the cortex via tangential migrations (Anderson et al 1997, Lavdas et al 1999, Sussel et al 1999). This will be the subject of another chapter in this volume (Parnavelas et al 2000, this volume). Here, I wish to briefly discuss some evidence that the development of cortical and subcortical neurons are under distinct genetic controls. For instance, while cortical neurons express the TBR1 transcription factor, subcortically derived cells express the *Dlx* genes (Bulfone et al 1998). This is clearly seen in the olfactory bulb, where the projection neurons (mitral cells) express TBR1 and the interneurons express DLX1 and DLX2 (Bulfone et al 1998). Thus, mutation of *Tbr1* largely eliminates mitral cells and mutation of *Dlx1* and *Dlx2* eliminates GABA expression in the olfactory bulb (Bulfone et al 1998). The distinct genetic regulation of cortical and subcortical cells may in fact be responsible for our observation that cortically derived cells are generally glutaminergic and subcortically derived cells are generally GABAergic.

Cortical regionalization could control areal differences by regulating proliferation (Rakic 1995, Polleux et al 1997), laminar histology (e.g. the CA fields of the hippocampus have only one layer of projection neurons whereas the neocortex has multiple layers of projection neurons), and the properties of the radial glia and non-GABAergic neurons. Radial migration of cortical projection neurons would translate the positional information of the cortical ventricular zone to the cortical mantle. Thus, regionalization of the cortical mantle zone could thereby provide positional information that would direct the pathfinding of afferent axon tracts, and direct the migration and distribution of the subcortically derived interneurons.

Acknowledgements

This work was supported by the research grants to JLRR from: Nina Ireland, National Alliance for Research on Schizophrenia and Depression, Human Frontiers Science Program, National Institute for Neurological Diseases and Stroke. Grant numbers NS34661-01A1 and NIMH K02 MH01046-01.

References

Abdel-Majid RM, Leong WL, Schalkwyk LC et al 1998 Loss of adenylyl cyclase I activity disrupts patterning of mouse somatosensory cortex. Nat Genet 19:289–291

Acampora D, Avantaggiato V, Tuorto F, Simeone A 1997 Genetic control of brain morphogenesis through *Otx* gene dosage requirement. Development 124:3639–3650

Anchan RM, Drake DP, Haines CF, Gerwe EA, LaMantia AS 1997 Disruption of local retinoid-mediated gene expression accompanies abnormal development in the mammalian olfactory pathway. J Comp Neurol 379:171–184 (erratum: 1997 J Comp Neurol 384:321)

Anderson SA, Eisenstat D, Shi L, Rubenstein JLR 1997 Interneuron migration from basal forebrain: dependence on *Dlx* genes. Science 278:474–476

Bulfone A, Wang F, Hevner R et al 1998 An olfactory sensory map develops in the absence of normal projection neurons or GABAergic interneurons. Neuron 21:1273–1282

Cases O, Vitalis T, Seif I, De Maeyer E, Sotelo C, Gaspar P 1996 Lack of barrels in the somatosensory cortex of monoamine oxidase A-deficient mice: role of a serotonin excess during the critical period. Neuron 16:297–307

Chiang C, Litingtung Y, Lee E et al 1996 Cyclopia and defective axial patterning in mice lacking *Sonic hedgehog* gene function. Nature 383:407–413

Cohen-Tannoudji M, Babinet C, Wassef M 1994 Early determination of a mouse somatosensory cortex marker. Nature 31:460–463

Dou CL, Li S, Lai E 1999 Dual role of brain factor-1 in regulating growth and patterning of the cerebral hemispheres. Cereb Cortex 9:543–550

Fernández A, Pieau C, Repérant J, Boncinelli E, Wassef M 1998 Expression of the *Emx-1* and *Dlx-1* homeobox genes define three molecularly distinct domains in the telencephalon of mouse, chick, turtle and frog embryos: implications for the evolution of telencephalic subdivisions in amniotes. Development 125:2099–2111

Furuta Y, Piston DW, Hogan BL 1997 Bone morphogenetic proteins (BMPs) as regulators of dorsal forebrain development. Development 124:2203–2212

Galceran J, Miyashita-Lin E, Devaney E, Rubenstein JLR, Grosschedl R 2000 Hippocampus development and generation of dentate gyrus granule cells is regulated by LEF-1. Development 127:469–482

Gitton Y, Cohen-Tannoudji M, Wassef M 1999 Role of thalamic axons in the expression of H-2ZI, a mouse somatosensory cortex specific marker. Cereb Cortex 9:611–620

Golden JA, Bracilovic A, McFadden KA, Beesely JS, Rubenstein JLR, Grinspan JB 1999 Ectopic bone morphogenetic proteins 5 and 4 in the chick forebrain leads to cyclopia and holoprosencephaly. Proc Natl Acad Sci USA 96:2439–2444

Grove EA, Tole S, Limon J, Yip L, Ragsdale CW 1998 The hem of the embryonic cerebral cortex is defined by the expression of multiple *Wnt* genes and is compromised in Gli3-deficient mice. Development 125:2315–2325

Hubel DH 1988 Eye, brain and vision. Scientific American Library, New York

Ikeya M, Lee SM, Johnson JE, McMahon AP, Takada S 1997 Wnt signalling required for expansion of neural crest and CNS progenitors. Nature 389:966–970

Innocenti G 1995 Exuberant development of connections, and its possible permissive role in cortical evolution. Trends Neurosci 18:397–402

Iwasato T, Erzurumlu RS, Huerta PT et al 1997 NMDA receptor-dependent refinement of somatotopic maps. Neuron 19:1201–1210

Killackey HP, Rhoades RW, Bennett-Clarke CA 1995 The formation of a somatotopic map. Trends Neurosci 18:402–407

Kohtz JD, Baker DP, Corte G, Fishell G 1998 Regionalization within the mammalian telencephalon is mediated by changes in responsiveness to Sonic Hedgehog. Development 125:5079–5089

Krubitzer L 1995 The organization of neocortex in mammals: are species differences really so different? Trends Neurosci 18:408–417

Lavdas AA, Grigoriou M, Pachnis V, Parnavelas JG 1999 The medial ganglionic eminence gives rise to a population of early neurons in the developing cerebral cortex. J Neurosci 19:7881–7888

Lee KJ, Mendelsohn M, Jessell TM 1998 Neuronal patterning by BMPs: a requirement for GDF7 in the generation of a discrete class of commissural interneurons in the mouse spinal cord. Genes Dev 12:3394–3407

Levitt P, Barbe MF, Eagleson KL 1997 Patterning and specification of the cerebral cortex. Annu Rev Neurosci 20:1–24

Martinez S, Crossley PH, Cobos I, Rubenstein JLR, Martin GR 1999 FGF8 induces formation of an ectopic isthmic organizer and isthmocerebellar development via a repressive effect on *Otx2* expression. Development 126:1189–1200

Miyashita-Lin EM, Hevner R, Montzka Wassarman K, Martinez S, Rubenstein JLR 1999 Early neocortical regionalization is preserved in the absence of thalamic innervation. Science 285:906–909

Northcutt RG, Kaas JH 1995 The emergence and evolution of mammalian neocortex. Trends Neurosci 18:373–379

Nothias F, Fishell G, Ruiz i Altaba A 1998 Cooperation of intrinsic and extrinsic signals in the elaboration of regional identity in the posterior cerebral cortex. Curr Biol 8:459–462

Parnavelas JG, Anderson SA, Lavdas AA, Grigoriou M, Panchis V, Rubenstein JLR 2000 The contribution of the ganglionic eminence to the neuronal cell types of the cerebral cortex. In: Evolutionary developmental biology of the cerebral cortex. Wiley, Chichester (Novartis Found Symp 228) p 129–147

Parr BA, Shea MJ, Vassileva G, McMahon AP 1993 Mouse *Wnt* genes exhibit discrete domains of expression in the early embryonic CNS and limb buds. Development 119:247–261

Pellegrini M, Mansouri A, Simeone A, Boncinelli E, Gruss P 1996 Dentate gyrus formation requires Emx2. Development 122:3893–3898

Polleux F, Dehay C, Moraillon B, Kennedy H 1997 Regulation of neuroblast cell-cycle kinetics plays a crucial role in the generation of unique features of neocortical areas. J Neurosci 17:7763–7783

Puelles L, Kuwana E, Bulfone A et al 2000 Pallial and subpallial derivatives in the embryonic chick and mouse telencephalon, traced by the expression of the *Dlx-2, Emx-1, Nkx-2.1, Pax-6* and *Tbr-1* genes. submitted

Qiu M, Anderson S, Chen S et al 1996 Mutation of the *Emx-1* homeobox gene disrupts the corpus callosum. Dev Biol 178:174–178

Rakic P 1988 Specification of cerebral cortical areas. Science 241:170–176

Rakic P 1995 A small step for the cell, a giant leap for mankind: a hypothesis of neocortical expansion during evolution. Trends Neurosci 18:383–388

Rubenstein JLR, Beachy PA 1998 Patterning of the embryonic forebrain. Curr Opin Neurobiol 8:18–26

Rubenstein JLR, Shimamura K, Martinez S, Puelles L 1998 Regionalization of the prosencephalic neural plate. Annu Rev Neurosci 21:445–477

Rubenstein JLR, Anderson SA, Shi L, Miyashita-Lin E, Bulfone A, Hevner R 1999 Genetic control of cortical regionalization and connectivity. Cereb Cortex 9:524–532

Schlagger BL, O'Leary DDM 1991 Potential of visual cortex to develop an array of functional units unique to somatosensory cortex. Science 252:1556–1560

Shimamura K, Rubenstein JLR 1997 Inductive interactions direct early regionalization of the mouse forebrain. Development 124:2709–2718

Shimamura K, Hartigan DJ, Martinez S, Puelles L, Rubenstein JLR 1995 Longitudinal organization of the anterior neural plate and neural tube. Development 121:3923–3933

Sussel L, Kimura S, Rubenstein JLR 1999 Loss of *Nkx2.1* homeobox gene function results in a ventral to dorsal respecification within the basal telencephalon: evidence for a transformation of the pallidum into the striatum. Development 126:3359–3370

Tan SS, Kalloniatis M, Sturm K, Tam PP, Reese BE, Faulkner-Jones B 1998 Separate progenitors for radial and tangential cell dispersion during development of the cerebral neocortex. Neuron 21:295–304

Tole S, Grove EA 1999 Patterning events and specification signals in the developing hippocampus. Cereb Cortex 9:551–561

Tole S, Christian C, Grove EA 1997 Early specification and autonomous development of cortical fields in the mouse hippocampus. Development 124:4959–4970

Welker E, Armstrong-James M, Bronchti G et al 1996 Altered sensory processing in the somatosensory cortex of the mouse mutant *barrelless*. Science 271:1864–1867

Wise SP, Jones EG 1978 Developmental studies of thalamocortical and commissural connections in the rat somatic sensory cortex. J Comp Neurol 178:187–208

Xuan S, Baptista CA, Balas G, Tao W, Soares VC, Lai E 1995 Winged helix transcription factor BF-1 is essential for the development of the cerebral hemispheres. Neuron 14:1141–1152

Ye W, Shimamura K, Rubenstein JLR, Hynes M, Rosenthal A 1998 Intersections of the FGF8 and Shh signals create inductive centers for dopaminergic and serotonin neurons in the anterior neural plate. Cell 93:755–766

Yoshida M, Suda Y, Matsuo I et al 1997 Emx1 and Emx2 functions in development of dorsal telencephalon. Development 124:101–111

DISCUSSION

Purves: It is a shame that the *Gbx2* mutant dies at birth. Is there any chance of carrying it longer in order to observe what might happen later?

Rubenstein: I don't know why it dies. It has defects in the hindbrain, cerebellum and craniofacial structures, as well as defects in other parts of the body. We are trying various strategies to keep it alive.

Karten: If you made slices, how long could you keep those going? Because this might be a useful strategy for looking at lethal mutations of the cortex.

Purves: For days, or even weeks if the slices are cultured.

Rubenstein: I would still prefer to do these studies *in vivo*.

Reiner: I have question concerning the mechanisms of cortical outgrowth in general. Do the cortical fibres reach the thalamic fibres, or do they fail to reach each other?

Rubenstein: Robert Hevner has done this work (Hevner et al 2000). They both grow into the internal capsule, but they don't seem to get close. We haven't measured how close they get.

Kaas: I have question about your marker for somatosensory cortex.

Rubenstein: We don't have one. We used Id2 to distinguish the boundary between somatosensory and non-somatosensory areas, but it is not a unique marker of somatosensory cortex because it also enters the visual cortex.

Kaas: That was my question. Most markers of somatosensory cortex also mark auditory and visual areas.

Rubenstein: This does as well. This assay (see Rubenstein et al 1999, Miyashita-Lin et al 1999) just tells us that you don't need thalamic axons to make the motor–sensory boundary.

Molnár: Did you relate the early, postnatal day (P) 0 gene expression pattern to the ones observed at later stages in the wild-type to show that this boundary is indeed the boundary of the motor cortex? Would the gene expression work in an adult, in which you could correlate it to cytoarchitectonics and then work backwards to P0? It is important to show that the early gene expression at birth is indeed related to the future cytoarchitectonics in adult, where one has a larger repertoire of markers.

Rubenstein: The use of histochemical staining for cytochrome oxidase and serotonin at P6 is a reliable way to identify the barrel fields. These methods have been used to define neocortical subdivisions (Killackey et al 1995). Therefore, in my opinion, our labelling analysis of Id2 at P6 is sufficient to say that this is the rostral end of the somatosensory cortex. I haven't been convinced by any other arguments so far, so is it the rostral end of the somatosensory cortex?

Karten: How would you go about deciding this?

Rubenstein: To do it properly, you would need to do a fate-mapping analysis, i.e. you would have to mark the cells at the position at the rostral end of the serotonin and Id2 staining at P6, and follow them over time to determine where they map in the adult.

Kaas: I would like to see whether this pattern can be reproduced in the visual cortex, although it may be quite crude at P1.

Rubenstein: It is unlikely that I can prove that at P0 we are looking at the transition between sensory cortex and motor cortex. However, the point is that a molecule marker which approximates this transition is already present and is unperturbed in the absence of thalamic input.

Karten: So, in other words there is a distinct regionalization, and we all agree that there are different cortical areas. What genes might regulate such an event?

Molnár: Genes triggering regionalization could still be activated through external connectivity in the *Gbx2* knockout mouse. One connection is maintained, i.e. the layer V projections through the cerebral peduncle to the spinal cord and other subcortical structures. Perhaps there are differences in these projections between the rostral and dorsal cortical areas, in which case in theory, they could retrogradely trigger regionalization and they wouldn't need thalamic projections.

Karten: Dennis O'Leary has looked at when the various efferents are established and to what extent they are specifying the cortex from which they are arising. Is it likely that they could give such a signal before they have specified, or do these signals arise after the cortical area is specified?

O'Leary: I can't imagine a scenario where layer V output axons are involved in specifying cortex *per se*, and in rodents they do not exhibit mature areal specificity until the second or third postnatal week.

Rakic: I agree, but even if they do, then you would have to ask what is telling layer V cells to project to a given subcortical target; so you would have to say that intrinsic genetic programmes are involved at this level.

Rubenstein: We have to think about the mechanisms controlling regionalization within the ventricular zone of the pallium, and about the patterning centres. We have to find the regulatory genes that are expressed in those areas and then manipulate them.

Karten: Cytoarchitectonic areas have baffled people since they were first described over a 100 years ago because they are complex assemblages of heterogeneous groups of neurons, and yet they're absolutely identifiable. What is regulating such an event? What second messenger systems are involved?

Rakic: I would like to make a further point of clarification, i.e. that we need to define more precisely what we mean by the term 'area'. For example, most people consider that the border between area 17 and 18 defines the visual cortex, while the other may think that it is between areas 18 and 19 or area 19 and 20. In fact Van Essen would consider most of the cerebral cortex to be the visual (Felleman & Van Essen 1991). Therefore, when we talk about the definition of cytoarchitectonic areas in developing primates, we're not talking about 100 different fields. There may have originally been only seven larger primary fields, which are than subsequently divided into smaller subfields. The pattern of thalamic input and functional analysis by recording or optical imaging would then enable dividing the areas further. However, these subdivisions belong to a different category.

Karten: But can we think in terms of thalamic inputs? In your *Gbx2* knockouts, there is some degree of cortical differentiation, in terms of regions as well as cortical laminae, in the absence of any thalamic input, so something else must be responsible for generating these cortical areas.

O'Leary: Presumably, regulatory genes differentially expressed across the cortex, for example in gradients, specify cortical area identities in the absence of thalamic input. Candidates include *Emx2* and *Lxh2*, which are expressed in a high caudal to low rostral gradient across the embryonic neocortex, and *Pax6* which is expressed in a countergradient. These genes may regulate a cascade that would control among other things the development of area-specific thalamocortical connections.

Karten: Do we have any strong evidence that these genes are expressed in gradients?

O'Leary: One problem is that the size of the cortex in mouse knockouts of these genes is reduced. This suggests that if these genes do have roles in specifying position, they also have other roles.

Herrup: Are we sure that the regionalization code is only carried by the neurons, or could it be carried by non-neuronal cells, e.g. the radial units?

Goffinet: If it is being carried by non-neuronal cells, then we have to envisage that diffusible factors are involved.

Herrup: And you need cell contact, but there's plenty of that between neurons and non-neuronal cells.

Goffinet: The relationship between the transcription factor code and diffusible factors is not clear to me. Are some of them down-regulated by bone morphogenetic proteins (BMPs)?

Rubenstein: There's evidence that BMPs up-regulate dorsal genes and down-regulate ventral genes (Golden et al 1999). People are working on the patterning roles of these secreted morphogens and on their ability to regulate the expression of transcription factors that are essential for cell fate determination and regionalization. It's a little early to say anything at this stage, but I suspect that within the next year or so there will be some interesting papers published on this.

O'Leary: It is interesting that the arealization of the neocortex appears to relate to the more global regionalization of the pallium itself, because many genes that have graded expression patterns across the dorsal telencephalon are expressed at high concentrations in the hippocampus, and their expression decreases in a graded fashion through the neocortex, or vice versa.

Rubenstein: The next level of organizational principles involves the secondary higher association cortices. This process may be somewhat independent of the patterning programme within the ventricular zone.

Puelles: I would like to point out that there is another level of organization that involves secreted proteins, and particularly the cadherin protein family, in which there is a distinct relationship between the different cadherins and particular cortical fields (Redies 1997, Suzuki et al 1997). Differential cadherin expression appears early in the ventricular zone and then appears in the radial glia and migrated neurons. Therefore, there could be a mechanism that transfers specification in the ventricular zone to other levels of histogenetic activity.

Levitt: What could the patterning programmes be coding, if it's not coding the information necessary to allow the cortical domains to connect in different ways?

Karten: Has it been shown that cadherin is doing this? What we've shown are coincident distributions of cadherins that match patternings.

Levitt: We did transplant studies in the 1990s in which we changed the expression patterns of an axon guidance molecule, and observed changes in the thalamocortical and corticocortical relationships within that piece of tissue (Barbe & Levitt 1992, 1995).

O'Leary: We can also take some lessons from the visual system, where it's been clearly demonstrated that ectopic expression of *engrailed* genes, which are normally expressed in a graded pattern in the dorsal midbrain, induces the expression of the axon guidance molecules, ephrin A2 and A5. This ectopic expression alters the normal topography of the retinotectal map. The same things likely occur in the cortex.

Levitt: It must be coding positional information that is ultimately used for generating connections.

Rubenstein: *Gbx2* mutants lack serotonin-containing thalamic afferents to the cortex. The median forebrain bundle from the raphe is there, but it's unlikely that this has anything to do with regionalization because its fibres are dispersed throughout the cortex. I don't have any information about the basal forebrain cholinergic system, but it is formed relatively late in development.

Welker: Is it possible that in the absence of the thalamus, axons from the trigeminal complex in the brainstem enter directly into the cortex?

Rubenstein: It's unlikely because the trigeminal nucleus forms in rhombomeres 1, 2 and 3, and those are the regions that are almost completely destroyed in the *Gbx2* mutant.

I also expected the raphe to disappear, but this is not the case. I was surprised that serotonergic neurons were present in rhombomeres 1 and 2.

Levitt: You mentioned the concept of the organizing centre that might organize topology or other positional information within the cortex. It is possible that this could be derived from the outside. Have you looked for mutations of the basal ganglia, because this region may be a logical candidate to serve as an organizing centre for the pallium?

Rubenstein: We wondered whether expression of *sonic hedgehog* (*Shh*) at the base of the telencephalon may be involved in global regionalization within the telencephalon. In the *Nkx2.1* mutant, *Shh* is no longer expressed in this region, suggesting that Nkx2.1 is upstream of Shh in the basal telencephalon. As far as we can tell, the cortex is normal in the absence of Shh in the telencephalon, except for a reduction of interneurons, suggesting that *Shh* expression within the telencephalon is probably not important. Many groups are working on dorsal patterning mutants (e.g. mutants of *Wnts* and *Bmps*), so within the next year or two we should know more about the effects of mutating genes involved in dorsal patterning.

Wolpert: What sort of distances are you talking about when you refer to organizing centres? Isn't the diffusion of these molecules all over by then?

Rubenstein: In the neural plate, the region I'm calling the rostral organizing centre is defined by the expression of fibroblast growth factor (FGF)8. It is located in the anterior neural ridge (ANR), which encompasses much of the anlage of the telencephalon. However, like a growth zone within a limb, there is expansion, so we like to think of the ANR as being similar to the apical ectodermal ridge (AER).

Molnár: What is the relationship between the thalamocortical and the early corticofugal axons to the Tbr1-positive/Emx1-positive and Tbr1-positive/Emx1-negative stripe of tissue at the striato–cortical boundary? Although I'm one of the proposers of the handshake hypothesis (Molnár & Blakemore 1995), there are many other possible explanations. Perhaps the stripe of tissue at the striato-cortical junction is thicker so the thalamocortical and early corticofugal projections cannot penetrate it to continue their journey.

Rubenstein: The corticofugal axons in the *Tbr1* mutants penetrate through the external capsule and enter the striatum, (passing through the Tbr1-positive/Emx1-negative zone), suggesting that defects in this pallial domain are not the major reason why the corticothalamic and thalamocortical pathways don't form in this mutant.

Puelles: It would be apposite to mention here what happens to the striatopallial boundary in the *Pax6* mutant.

Rubenstein: This is the work of Peter Gruss' laboratory, who found that in the *Pax6* mutant, there seems to be a 'breaking of the dam' at the boundary between the subpallium and the pallium, apparently leading to an increased number of

Dlx-positive cells in the cortex (Stoykova et al 1996). This suggests that there's a decreased resistance to the subpallial migration to the cortex. There may also be patterning effects because Pax6 has been known to repress *Nkx* genes, as Tom Jessell has shown. Pax6 represses *Nkx2.2* in the spinal cord (Ericson et al 1997). Therefore, by reducing *Pax6* expression, you may also ventralize the telencephalon, so some of this apparent migration may actually be respecification of the proliferative zone, i.e. ventral pallium regions into striatal regions.

Levitt: Gruss and colleagues have also showed that there are severe radial glial alterations, and that the pallidal-subpallidal junction is also altered (Götz et al 1998).

Rakic: Much of this work has been performed, by necessity, in mice. I would like to point out, however, that these expression patterns are also conserved in humans and non-human primates. For example, Maria Donoghue and I examined the expression patterns of about 12 different genes, including the eph receptor and ephrins, and found essentially the same patterns as in the mouse, except that the gradients were even sharper because monkey cortex has 100-fold larger surface. It is interesting that some of these genes were expressed in the tectum where they form similar gradients and are thought to be involved in the formation of topographic maps.

Kaas: But there are many more subdivisions in humans and non-human primates than in the mouse, so these additional subunits cannot be explained the universal gradients.

Rakic: I agree, but I don't know whether the subdivisions between areas in the human frontal lobe are the same as the border between 17 and 18 in the monkey and mouse. This is why we first have to define areas and find out whether those subdivisions have similar interactions with corresponding thalamic nuclei.

Puelles: This is analogous to the fate of empires, in the sense that, once their size increases past a given threshold, then direct communication between the elements becomes weaker and a tendency to desegregate into independent units is observed. There are lateral inhibitory effects in neuroepithelium, as in human societies; each ventricular cell is affected in such a way that its full developmental potential is not expressed. In the cortex, you may likewise need an expanded area in order to favour the emergence of novel areas, and this might be related to quantitative dynamic changes in the lateral interactions among prespecified neuroepithelial cells.

Karten: Another issue in regard to the area and regionalization problem is that some of the laminae which share common gene expression patterns overlap with several different cortical areas, whereas others terminate abruptly. In this sense, cortex is much more of a laminar structure, whereas from a functional point of view, it is a radial structure. Is this something we should be concerned about?

Rakic: No. Cortex is both a laminar and radial structure because space is three-dimensional.

Karten: You have also suggested that new cortical zones are generated by a mechanism that increases the surface area. But as the surface area increases, presumably areas with different relative representations are made, and these sometimes lack certain sublayers that express certain genes. Can these two points be reconciled?

Herrup: No. The model we heard from Lewis Wolpert first predicts that this would happen. As the distances increase and the gradients stretch, areas with shallow gradients appear, and in these regions there are opportunities for new combinations of factors to act on the enhancers.

O'Leary: In addition, there's evidence from Pasko Rakic's lab that thalamic input plays a major role in the differentiation of area-specific cytoarchitecture.

Rakic: The cells have to be in the correct position establish appropriate connections. If they are not in the correct position, e.g. if the geniculate nucleus projects to the frontal lobe, their axons may not encounter the correct receptors and so they would not generate an appropriate response. For example, in the embryonic human forebrain, there must be genes that induce the production of additional cells to accommodate for the increased number of neurons in layer IV of the visual cortex. This cannot fully develop without interactions with the thalamus, because when we diminish the size of the geniculate input by 30%, we see that area 17 shrinks by 30% (Rakic 1988). Adjacent to this is what I call hybrid cortex because it has some molecular properties of area 17, but comes from area 18 (Rakic et al 1991).

Reiner: The idea that gradients of molecules are involved in specifying cortical areas is plausible and interesting, but it seems to me to be a fairly imprecise way of specifying areas. If it is really true, I would expect there to be variation within a species with respect to the extent of individual cortical areas.

Rubenstein: In his presentation, Lewis Wolpert described gradients of two transcription factors in the *Drosophila* embryo. These gradients can be read by enhancer elements of the specifying genes to give rise to razor-sharp stripes. In the developing sensory cortex we're beginning to hypothesize the existence of a similar phenomenon. For instance, Dennis O'Leary's group has pointed out gradients of transcription factor expression, such as Emx2 gradients. In addition, some eph receptors appear to be expressed in discrete domains — these molecules are implicated in regulating the topographic map of synaptic inputs.

Molnár: This, however, doesn't exclude the possibility that some of these gradients can be redefined. For instance, if the occipital cortical neuroepithelial sheet were removed at very early embryonic stages, perhaps the gradients in the early gene expression patterns were redefined and a different part of the cortical neuroepithelium would give rise to visual cortex. This suggests that there is a distinction between the proto-map hypothesis and early gene expression gradients in the cortex.

Rakic: No. A 'proto-map' is not the same as a 'fate map'. 'Proto' is a general term for a malleable primordium, which means that a 'proto-map' can be modified. The border between cortical areas can then be defined by the intersection of different concentration gradients that can be sharpened by thalamic inputs and reciprocal interactions between neurons. When Brodmann (1909) was defining his cytoarchitectonic maps, he was looking for morphological expression, whereas we are looking at how this morphological expression is set up, and we believe that although modifiable, it is genetically constrained.

References

Barbe MF, Levitt P 1992 Attraction of specific thalamic input by cerebral grafts depends on the molecular identity of the implant. Proc Natl Acad Sci USA 89:3706–3710

Barbe MF, Levitt P 1995 Age-dependent specification of the corticocortical connections of cerebral grafts. J Neurosci 15:1819–1834

Brodmann K 1909 Vergleichende Localisationsationslehre der Grosshir-hinde. Barth, Leipzig

Ericson J, Rashbass P, Schedl A et al 1997 Pax6 controls progenitor cell identity and neuronal fate in response to graded Shh signalling. Cell 90:169–180

Felleman DJ, Van Essen DC 1991 Distributed hierarchical processing in the primate cerebral cortex. Cereb Cortex 1:1–47

Golden JA, Bracilovic A, McFadden KA, Beesely JS, Rubenstein JLR, Grinspan JB 1999 Ectopic bone morphogenetic proteins 5 and 4 in the chick forebrain leads to cyclopia and holoprosencephaly. Proc Natl Acad Sci USA 96:2439–2444

Götz M, Stoykova A, Gruss P 1998 Pax6 controls radial glia differentiation in the cerebral cortex. Neuron 21:1031–1044

Hevner RF, Miyashita-Lin E, Marin O, Rubenstein JLR 2000 Coordination of cortical and thalamic axon pathfinding, submitted

Killackey HP, Rhoades RW, Bennett-Clarke CA 1995 The formation of a somatotopic map. Trends Neurosci 18:402–407

Miyashita-Lin EM, Hevner R, Montzka Wassarman K, Martinez S, Rubenstein JLR 1999 Early neocortical regionalization is preserved in the absence of thalamic innervation. Science 285:906–909

Molnár Z, Blakemore C 1995 How do thalamic neurons fond their way to the cortex? Trends Neurosci 18:389–397

Rakic P 1988 Specification of cerebral cortical areas. Science 241:170–176

Rakic P, Suñer I, Williams RW 1991 A novel cytoarchitectonic area induced experimentally within the primate visual cortex. Proc Natl Acad Sci USA 88:2083–2087

Redies C 1997 Cadherins and the formation of neural circuitry in the vertebrate CNS. Cell Tissue Res 290:405–413

Rubenstein JLR, Anderson SA, Shi L, Miyashita-Lin E, Bulfone A, Hevner R 1999 Genetic control of cortical regionalization and connectivity. Cereb Cortex 9:524–532

Stoykova A, Fritsch R, Walther C, Gruss P 1996 Forebrain patterning defects in *Smalleye* mutant mice. Development 122:3453–3465

Suzuki SC, Inoue T, Kimura Y, Tanaka T, Takeichi M 1997 Neuronal circuits are subdivided by differential expression of type-II classic cadherins in postnatal mouse brains. Mol Cell Neurosci 9:433–447

A hypothesis as to the organization of cerebral cortex in the common amniote ancestor of modern reptiles and mammals

Anton J. Reiner

Department of Anatomy and Neurobiology, College of Medicine, University of Tennessee, 855 Monroe Avenue, Memphis, TN 38163, USA

Abstract. Opinions on the evolutionary origins of mammalian neocortex have divided into two camps: (1) antecedents of the superior neocortex (i.e. occipital, parietal and frontal lobes) and temporal neocortex (i.e. temporal lobe) were present in stem amniotes, and these antecedent regions gave rise to dorsal cortex and dorsal ventricular ridge (DVR), respectively, in living reptiles; (2) the stem amniote antecedent of mammalian superior neocortex gave rise to dorsal cortex in the reptilian lineage, while the stem amniote antecedent of mammal claustrum, endopiriform region and/or basolateral/basomedial amygdala gave rise to DVR in reptiles, with mammalian temporal neocortex being a newly evolved structure with no reptilian homologue. The latter hypothesis has the merit of being more consistent with some current homeobox gene data, but it has the disadvantages of positing that mammalian temporal neocortex arose *de novo*, and of assuming that the high similarity between DVR and temporal neocortex in the organization of thalamic sensory input and corticostriatal projections and in the topology of sensory areas is coincidental. If one assumes that the antecedent of superior and temporal neocortex in stem amniotes was one continuous field that histologically resembled dorsal cortex in living reptiles, the first hypothesis provides basis for a parsimonious account of the origin of superior and temporal neocortex and their considerable resemblance to dorsal cortex and DVR in reptiles, as well as to Wulst and DVR in birds.

2000 Evolutionary developmental biology of the cerebral cortex. Wiley, Chichester (Novartis Foundation Symposium 228) p 83–108

Basic questions about the evolution of neocortex in mammals

The neocortex is the portion of the telencephalon in mammals that is thought to underlie the higher order perceptual, cognitive and learning abilities of mammals (Reiner et al 1984, Jerison 1985, Allman 1990, Arbib et al 1998). For this reason, considerable attention has been devoted to the evolutionary origins of neocortex and on how the evolution of neocortex might explain any demonstrable differences between mammals and non-mammals in perceptual, cognitive and learning capacity (Hodos 1982, Macphail 1982, Northcutt & Kaas 1995). Questions about the evolutionary origins of neocortex can be broken down into at least three sub-questions. Since mammals evolved from stem amniotes, these questions need to focus on the stem amniote–mammal transition. First, at what point in this transition did neocortex arise? This question is easily answered. No living non-mammal has a telencephalic structure that possesses the six-layered cytoarchitecture characteristic of mammalian neocortex, and all living mammals do (Allman 1990, Northcutt & Kaas 1995). Thus, neocortex evolved uniquely in the mammalian lineage after its divergence from the lineage leading to living reptiles but before the radiation of mammals into monotremes, marsupials and placentals. Second, was neocortex derived from any antecedent structures within the telencephalon of stem amniotes, and do living reptiles possess their own unique derivatives of those same antecedent structures? If this question is answered in the affirmative, it raises the related question: what was the antecedent organization of the telencephalon in stem amniotes? This issue has not been resolved, but opinions have recently divided into two camps. One of these views I shall term the 'temporal neocortex *de novo* hypothesis' (Bruce & Neary 1995, Striedter 1997, Fernández et al 1998, Puelles et al 2000), while the other I shall refer to as the 'common origin of the temporal neocortex and dorsal ventricular ridge (DVR) hypothesis' (Butler 1994a,b, Karten 1969, 1991, Nauta & Karten 1970, Reiner 1993). It is one purpose of this review to discuss the merits of these two positions (Fig. 1), and to evaluate the relative parsimony with which they can explain both neocortical evolution and the evolution of what are certainly at least the functionally analogous telencephalic areas in living reptiles and birds. Thirdly, what transformations were necessary to evolve mammalian neocortex from whatever may have been the antecedent organization in stem amniotes? An answer to this question naturally requires knowing or assuming what the antecedent organization was in stem amniotes. As part of this overview, I will discuss the transformations in the organization of the telencephalon in stem amniotes that must be assumed by each of the two current hypotheses about the evolutionary origins of mammalian neocortex.

From what did neocortex in mammals evolve?

Mammalian neocortex is a six-layered structure, with the cortical layers disposed parallel to the cortical pial surface, and concentric to the centre of the hemisphere (Jones 1981). The neocortical layers are specialized in function, with layer IV (the granular layer) being the major target of specific sensory or motor input from the dorsal thalamus, layers II and III giving rise to corticocortical efferent projections, and layers V and VI giving rise to projections outside the neocortex (Jones 1981).

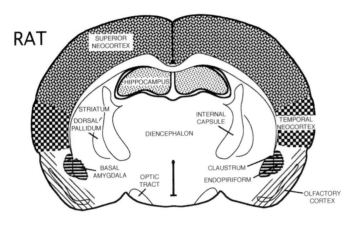

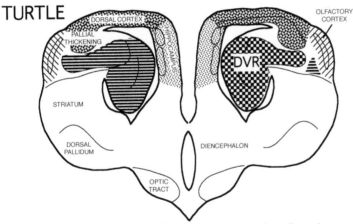

FIG. 1.   Proposed pallial homologies between mammals and reptiles, using transverse sections through rat telencephalon and turtle telencephalon for illustration, according to the 'temporal neocortex *de novo* hypothesis' and the 'common origin of temporal neocortex and dorsal ventricular ridge (DVR) hypothesis'. The key difference between the two hypotheses concerns the homology of the DVR of reptiles and birds. The *de novo* hypothesis proposes that temporal neocortex has no homologue in reptiles and birds, and that reptilian/avian DVR is homologous to the claustrum, endopiriform region and/or parts of the basolateral/basomedial amygdala. By contrast, the common origin hypothesis proposes that rostral DVR of reptiles and birds is homologous to temporal neocortex of mammals. While the reptilian/avian homologue of the claustrum–endopiriform–basolateral/basomedial amygdaloid complex is not entirely clear, it may include a small region just deep to olfactory cortex or to parts of caudal DVR. Shadings are used to highlight the homologous structures according to each of these two hypotheses.

The neurons of neocortex in the supra- and infragranular layers are characterized by having dendrites that span the cortical layers, which provides the basis for the functionally columnar organization of information processing within neocortex (Jones 1981). The neocortex varies in its extent among mammals, but in all cases it occupies most of the superior and temporal surfaces of the telencephalon, with the hippocampal complex abutting the medioventral edge of the neocortex, and the olfactory cortex, claustrum and endopiriform region abutting the lateroventral edge of the neocortex (Allman 1990). The neocortex appears to consist of two distinct zones that are demarcated by the temporal sulcus, which is evident (albeit shallow) even in mammals such as rodents that do not have a distinct temporal lobe (Arimatsu et al 1999). I shall refer to the neocortical field medial to the temporal sulcus as superior neocortex, and the part lateral to the temporal sulcus as temporal neocortex. Of interest for this chapter, the superior neocortex contains the primary visual area (V1, which receives its major visual input from the dorsal lateral geniculate nucleus of the thalamus) and the primary somatosensory area (S1, which receives its major sensory input from the ventrobasal thalamus). Similarly, it is of note that temporal neocortex contains the primary auditory cortex (A1, which receives its input from the medial geniculate nucleus), and the secondary visual area (V2, which we define here as the region receiving its major visual input from the lateral posterior/pulvinar region of the thalamus, which itself receives its visual input from the superficial layers of the superior colliculus) (Diamond & Hall 1969, Kaas 1980, Coleman & Clerici 1981, Rowe 1990, Northcutt & Kaas 1995, Beck et al 1996, Major et al 1998, Pobirsky et al 1998).

Both major schools of thought on neocortical evolution assume that stem amniotes possessed a hippocampal region within their telencephalic pallium and an olfactory (pyriform) cortex, but differ in their interpretation of how the neocortex arose. For an appreciation of how these two positions differ, it is necessary to briefly review telencephalic organization in living reptiles, since hypotheses as to neocortical evolution have been based on claims of telencephalic homology between extant reptiles and mammals. To simplify the discussion of the reptilian neocortical forerunner, I shall focus this chapter on turtles, since they are typically regarded as showing relatively primitive forebrain organization among reptiles, since more data are available for them than for other reptilian groups and since the major features of importance for this chapter appear similar in other reptilian groups (Orrego 1961, Hall et al 1977, Pritz 1980, Balaban & Ulinski 1981, Bruce & Butler 1984a,b, Ulinski 1988, Reiner 1993). Two portions of the pallium in modern turtles are relevant to the issue of the evolution of neocortex: (1) a cytoarchitecturally continuous cortical plate with a medial/dorsomedial hippocampal portion, and a more dorsally situated portion called the dorsal cortex; and (2) a subcortical dorsal ventricular ridge (DVR). The dorsal cortex shows a simple, three-layered organization, with a broad superficial cell-poor layer, a deep cellular layer and a relatively narrow subcellular layer (Fig. 1). The rostral part of dorsal cortex possesses a lateral extension that abuts the rostral part of the DVR, termed the pallial thickening (Johnston 1915, Reiner 1993). The layers of the reptilian dorsal cortex and pallial thickening are disposed parallel to the pial surface. By topology and connections, the dorsal cortex and pallial thickening of turtles resemble at least the superior part of mammalian neocortex (Reiner 1993, Butler 1994a, Striedter 1997). In addition, the lateral part of turtle dorsal cortex and the pallial thickening together contain a primary visual area (V1-like area) and an S1-like area (Orrego 1961, Hall et al 1977, Ulinski 1988, 1990, Reiner 1993). As is true of mammalian neocortex, dorsal cortex/pallial thickening shows a laminar segregation of thalamic input and cortical outputs, with the thalamic input ending in the superficial plexiform layer and the cortical output arising from the cellular layer (Ulinski 1988, Reiner 1993). This cortical output targets striatum and brainstem, and thereby resembles the output of layer V/VI in mammals (Ebner 1976, Reiner 1993). The dorsal cortex/pallial thickening, however, lacks the neuronal types unique to layers II–IV in mammals, as defined by connectivity and neurotransmitter content (Ebner 1976, Reiner 1991, 1993). Rather than contact layer IV-type granule cells, thalamic axons in turtle dorsal cortex/pallial thickening contact the apical dendrites of the pyramidal neurons, the latter of which also occurs in mammals (Jones 1981).

The DVR is a subcortical portion of the pallium that bulges into the lateral ventricle and is characteristic of the telencephalon of all living reptiles and of birds (Karten 1969, Nauta & Karten 1970, Northcutt 1981). The DVR in turtles differs from the dorsal cortex in cytoarchitecture, in that it consists of cell groups rather than layers (Johnston 1915, Elliot-Smith 1919, Durward 1930). The neurons within these cell groups possess radially symmetrical dendritic trees confined to the cell group within which the parent perikaryon resides (Ulinski 1990). Of interest, the rostral DVR contains cell groups that resemble, in terms of thalamic inputs, the V2 and A1 of mammalian neocortex (Balaban & Ulinski 1981, Ulinski 1990, Reiner 1993, Butler 1994b). This is true in birds as well (Karten 1969, 1991). The neurotransmitter organization of DVR seems similar to that of dorsal cortex/pallial thickening, and so it seems that DVR of turtles also lacks neurons resembling those found in layers II/III of mammalian neocortex (Reiner 1991, 1993). Based on the evidence (discussed below) that the DVR arose as a cell plate resembling that of dorsal cortex, I also believe that DVR in turtles lacks the type of granule cells found in layer IV of mammalian neocortex, and that thalamorecipient DVR neurons in turtles are like layer V/VI neurons of neocortex.

Of the two schools of thought on the evolution of mammalian neocortex, both accept that stem amniotes possessed a structure resembling the dorsal cortex/pallial thickening of turtles. Both hypotheses propose that this region was the forerunner of the superior part of mammalian neocortex, and of course of the dorsal cortex/

pallial thickening in living reptiles, as well as of Wulst in birds. The two hypotheses diverge as to the origin of temporal neocortex. One viewpoint proposes that temporal neocortex had no antecedent in stem amniotes and arose *de novo* in the mammalian lineage after the divergence of the lineages leading to modern mammals on one hand and living reptiles and birds on the other (Bruce & Neary 1995, Striedter 1997, Fernández et al 1998, Puelles et al 2000). This hypothesis further proposes that the DVR of reptiles, which seems to resemble temporal neocortex in possessing a V2-like and an A1-like region, is derived from a subcortical pallial region in stem amniotes that in the mammalian lineage came to give rise to the claustrum, the endopiriform region, and/or the basolateral/ basomedial amygdala (Fig. 1). This viewpoint I here refer to as the 'temporal neocortex *de novo* hypothesis', and it has its antecedents in the work of Holmgren (1925). The alternative hypothesis proposes that a region in stem amniotes situated at the ventrolateral edge of dorsal cortex/pallial thickening gave rise to temporal neocortex in the mammalian lineage and to at least rostral DVR in the sauropsid (reptile and bird) lineage (Karten 1969, 1991, Nauta & Karten 1970, Reiner 1993, Butler 1994a,b; Fig. 1). I here call this the 'common origin of DVR and temporal neocortex hypothesis'. In the following sections, I will briefly review the evidence for and against each of these hypotheses, and then while giving a more detailed account of the evolutionary transformations required by each attempt to evaluate the parsimony of each.

The temporal neocortex *de novo* hypothesis

Three lines of evidence have been used to argue for the homology of DVR of reptiles and birds to the claustrum, the endopiriform region and/or the basolateral/basomedial amygdala of mammals. First, it has been noted that DVR lies deep to pyriform (olfactory) cortex in reptiles and birds (Fig. 1). Since the claustrum, the endopiriform region and/or the basolateral/basomedial amygdala all lie deep to a greater or lesser extent to olfactory bulb recipient cortices (Fig. 1), it has been argued that topological considerations favour a homology of DVR of birds and reptiles to claustrum, the endopiriform region and/or the basolateral/basomedial amygdala of mammals (Bruce & Neary 1995, Striedter 1997). In general, however, using the topological relationship of two structures to make an argument for homology is problematic, since developmental events are not so invariant as to produce consistent neighbour relationships among a pair of brain structures across lineages (Swanson & Petrovich 1998, Puelles et al 2000). A second line of evidence for the homology of DVR of birds and reptiles to claustrum, the endopiriform region and/or the basolateral/basomedial amygdala of mammals concerns a claimed similarity in embryological derivation from the lateral edge of the pallial proliferative zone during neurogenesis (Streidter 1997).

The data for this interpretation are, however, not unambiguous, and some investigators have reported that DVR of birds and reptiles have a different spatial and temporal embryonic derivation from the pallial proliferative zone than do claustrum, the endopiriform region and the basolateral/basomedial amygdala of mammals (Källén 1951, 1962). The third and strongest line of evidence for the homology of DVR to subcortical pallial regions in mammals comes from recent homeobox gene-mapping studies (Fernández et al 1998, Puelles et al 2000). Several homeobox genes have been identified that are preferentially expressed in pallial regions, such as *Emx1*, *Emx2*, *Pax6* and *Tbr1* (Puelles & Rubinstein 1995, Fernández et al 1998, Puelles et al 2000). Among the genes expressed preferentially in pallial regions, *Tbr1*, *Emx2* and *Pax6* are expressed throughout the entire pallium in mammals, including the hippocampal cortex, the neocortex, the olfactory cortex, the claustrum/endopiriform region and the basolateral/basomedial amygdala. In contrast, expression of the *Emx1* gene is restricted to the developing hippocampal cortex, neocortex, olfactory cortex and claustrum, but it is absent from the endopiriform region, and parts of the basolateral/basomedial amygdala of mammals. Recent data on the expression of these genes in the embryonic telencephalon of chick and turtle confirm that the dorsal cortex, pallial thickening, olfactory cortex and DVR in reptiles are pallial in nature, as are the hippocampal complex, Wulst, olfactory cortex and DVR in birds (Fernández et al 1998, Puelles et al 2000). Thus, this evidence confirms the conclusions of prior studies on the embryology, connections and histochemistry of these regions in reptiles and birds (Källén 1951, 1962, Reiner et al 1984, 1998). In particular, this evidence supports the view that dorsal cortex/pallial thickening is homologous to the superior neocortex, since they use similar genes for regulating regional development (Fernández et al 1998, Puelles et al 2000). Of note, however, is that much of the DVR in turtles and birds does not express *Emx1* during development. In this regard, much of DVR resembles the endopiriform region and parts of the basolateral/basomedial amygdala of mammals. These homeobox expression data provide the most compelling evidence, in the opinion of this author, against the homology of the DVR of reptiles and the temporal neocortex of mammals.

Common origin of dorsal ventricular ridge and temporal neocortex hypothesis

Two major lines of evidence have been used to argue for the homology of temporal neocortex and DVR. First, the rostral dorsal cortex, pallial thickening and DVR appear to form one continuous structure in reptiles. This similarity is especially impressive in the primitive lizard *Sphenodon punctatum*, which may show the ancestral cytoarchitectural pattern for the DVR. In this species, the DVR is not broken into separate cell groups (Elliot-Smith 1919, Durward 1930). Rather, the DVR consists of a cell plate that resembles the cell plate of the dorsal cortex (Fig. 2). The neurons of this cell plate, which is continuous with the cell plate of the dorsal cortex/pallial thickening at rostral telencephalic levels, extend their apical dendrites into the centre of the DVR, where they appear to receive thalamic input (R. G. Northcutt, personal communication 1999). The neurons of the DVR cell plate extend basal dendrites into a cell-free zone that separates them from the ependyma of the DVR. Thus, except for its involution, DVR cytoarchitecture in *Sphenodon* closely resembles that of three-layered dorsal cortex. The *Sphenodon* data

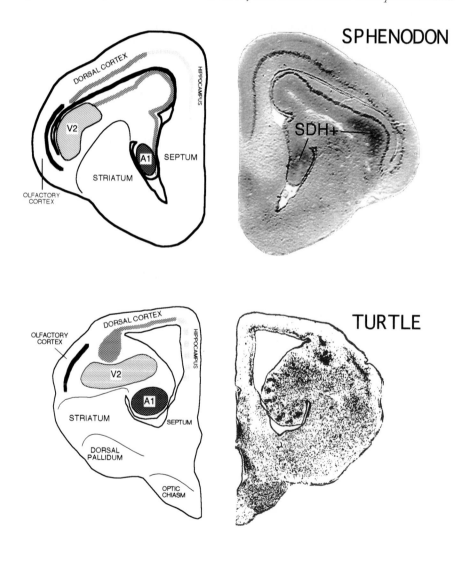

suggest that the dorsal cortex/pallial thickening and rostral DVR may have arisen as one continuous cortical cell field early in reptile evolution. The topological and cytoarchitectonic continuity of rostral dorsal cortex/pallial thickening and DVR in reptiles resembles that of superior and temporal neocortex in mammals, and has led some authors to propose the homology of rostral DVR and temporal neocortex (Reiner 1993, Butler 1994a,b). The second line of evidence for homology of rostral DVR and temporal neocortex deals with the high similarity in their connectivity. For example, both contain a V2 and an A1, with the thalamic and midbrain cell groups of origin for these inputs being remarkably similar in their neuroanatomical and functional organization between mammals on the one hand, and reptiles and birds on the other (Karten 1969, 1991, Pritz 1980, Balaban & Ulinski 1981, Bruce & Butler 1984a,b, Reiner 1993). While such similarities could have evolved independently, the argument that has been raised is that the similarities are so extreme in the midbrain and thalamic parts of the circuit that it is unlikely that they have evolved separately (Brauth & Reiner 1991, Reiner 1994, Luksh et al 1998, Major et al 1998). A third point that has not been raised previously but is noteworthy along these same lines concerns the topological arrangement of V1, V2, S1 and A1 in reptiles compared to primitive mammals likely to show the fundamental mammalian pattern (Fig. 3). Viewing the pallium from the side and with the DVR flattened, the neighbouring arrangement of these areas in turtles is nearly identical to that seen in a side-view section of the neocortex of primitive mammals (Orrego 1961, Hall et al 1977, Kaas 1980, Coleman & Clerici 1981, Rowe 1990, Reiner 1993, Beck et al 1996, Pobirsky et al 1998). This pattern could not have been a carry-over inheritance from the amphibian ancestors of stem amniotes, since there is no evidence from modern amphibians that these areas existed in ancestral amphibians (Northcutt & Kicliter 1980). Thus, the similarity

---

FIG. 2. (*opposite*) Photomicrographs and schematics illustrating and comparing the organization of *Sphenodon punctatum* and turtle telencephala. The illustrations for *Sphenodon* show an image of a transverse section through the rostral telencephalon that had been histochemically stained for succinic dehydrogenase (SDH) juxtaposed to a line drawing of this same section. The illustrations for turtle show a high contrast image of a cresyl violet-stained transverse section through the telencephalon of a painted turtle at the level of the anterior commissure juxtaposed to a line drawing of this same section. The major telencephalic subdivisions are identified, as are the secondary visual (V2, defined as receiving visual input via a retinotectothalamofugal pathway) and primary auditory (A1) areas of the rostral dorsal ventricular ridge (DVR). The dorsal cortex in *Sphenodon* and turtle constitutes a thin piece of tissue overlying the lateral ventricle, which grades into a distinct subcortical pallial thickening in turtles. Note that while the DVR itself in turtles consists of distinct cell groups and a few periventricular clusters of cells, in *Sphenodon* the DVR consists of a cell plate that histologically resembles the dorsal cortex. The two SDH$^+$ zones in *Sphenodon* DVR are likely to be V2 and A1, based on the documented efficacy of SDH histochemistry in identifying these sensory areas in reptile and bird DVR, and the position of these zones in birds and other reptilian species. The SDH-stained *Sphenodon* material was made available through the generosity of R. G. Northcutt.

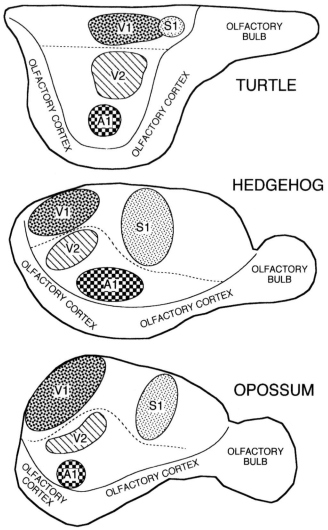

FIG. 3. Side-view line drawings illustrating the location and extent of the major thalamorecipient sensory areas in: (1) the turtle cortex and dorsal ventricular ridge (DVR), with DVR flattened to provide a schematized side view; (2) hedgehog neocortex; and (3) opossum neocortex. While many sensory areas are present in the neocortex of hedgehog and opossum, this figure focuses on those sensory areas found in turtle dorsal cortex and rostral DVR, namely V1, S1, V2 and A1. Note that V1 and S1 are above the dotted line, which separates dorsal cortex and DVR in turtles, and separates superior and temporal neocortex in mammals. By contrast, V2 and A1 are in the DVR of turtles and in the temporal neocortex of hedgehog and opossum. In addition note that S1 in all three cases is rostral to V1, while in all three cases V1, V2 and A1 are arrayed in a superior to inferior sequence. Placement of the sensory areas is based on the references cited in the text.

between modern reptiles and mammals in the topology of these 'cortical' sensory areas is likely to be due to common inheritance from a stem amniote possessing these same areas in this same configuration, since it seems too great to have occurred by coincidence.

Evolutionary transformations — the temporal neocortex *de novo* hypothesis

The *de novo* hypothesis must postulate a stem amniote common ancestor in which a dorsal cortex/pallial thickening was present, and which was presumably similar in cytoarchitecture to that in turtles (Fig. 4). In many respects, the morphology of such a region would also be likely to not differ much from that observed for the dorsal pallium in amphibians (Northcutt & Kicliter 1980). It seems likely that the dorsal cortex/pallial thickening of stem amniotes possessed a V1 and S1. The V1 may or may not have been inherited from ancestral amphibians — the hodological and electrophysiological data are not definitive on the presence of a V1 in the dorsal pallium of existing amphibians (Northcutt & Kicliter 1980). By the *de novo* hypothesis, the evolution of neocortex must have featured two major events: (1) a lateral expansion of the germinative epithelium giving rise to the neocortex, with this new region generating temporal neocortex; and (2) the later addition of layer II–IV cell types as products of neocortical histogenesis to all of neocortex (Fig. 4). Note that concomitant with this addition of new cell types presumably occurred a shift to an inside-out gradient of neurogenesis, away from the reptilian outside-in pattern (Marín-Padilla 1998). In the radiation leading to mammals, the small pallial region at the lateral edge of the neocortical zone that has been termed the intermediate zone and is devoid of *Emx1* expression is assumed to have given rise to such subcortical pallial structures as the endopiriform region and the basolateral/basomedial amygdala, with the claustrum arising from this or some neighbouring region (Bruce & Neary 1995, Striedter 1997, Fernández et al 1998, Puelles et al 2000). By contrast, in the radiation leading to living reptiles and birds, the *Emx1*-negative intermediate zone massively hypertrophied to become the DVR. In the opinion of this author, the main favourable points of parsimony to raise for this scenario are that it is consistent with the *Emx1* data and it explains why no clear-cut equivalents for the claustrum, the endopiriform region or the basolateral/basomedial amygdala, other than the DVR itself, are readily identifiable in reptilian pallium.

I believe, however, that this hypothesis is unparsimonious on several fronts. First, it posits a *de novo* origin for a major part of the mammalian neocortex. Secondly, it leaves to coincidence the considerable similarities in the topographic arrangement of sensory areas in the dorsal cortex/pallial thickening and DVR of turtles compared to that in primitive mammals that was noted above. Additionally, it leaves to coincidence the prominent similarities in the entire mesothalamocortical circuit for the A1 and V2 'cortical' fields of mammals and extant sauropsids (Brauth & Reiner 1991, Reiner 1994, Luksh et al 1998, Major et al 1998). One of the proponents of the *de novo* hypothesis has attempted to resolve this latter problem by suggesting that these circuits are, in fact, hypertrophied versions of the visual and auditory projections of intralaminar thalamus to mammalian amygdala (Bruce & Neary

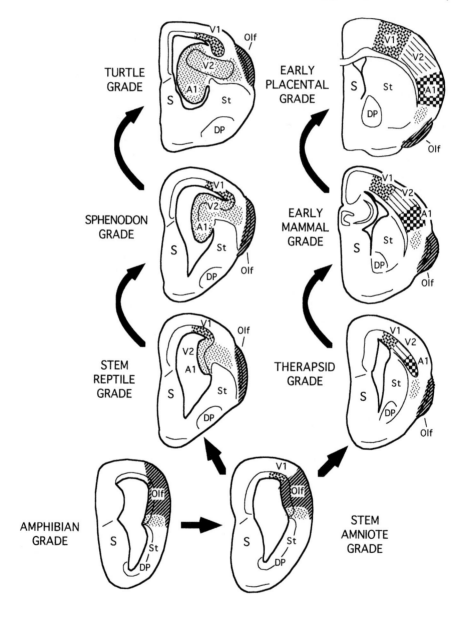

1995). Reptiles and birds, however, possess an intralaminar thalamus which resembles that in mammals, and this sauropsid intralaminar thalamus is not the source of the thalamic inputs to V2 and A1 in rostral DVR of reptiles and birds (Butler 1994a, Veenman et al 1997). Additionally, and as noted, the mesothalamo-V2 and mesothalamo-A1 circuits in reptiles and birds resemble the circuits to mammalian temporal neocortex, but not the visual and auditory intralaminar circuits to amygdala in mammals (Coleman & Clerici 1981, Brauth & Reiner 1991, Reiner 1994, Luksh et al 1998, Major et al 1998). Finally, in considering neocortical evolution, it is also important to consider the evolution of the basal ganglia as well. The basal ganglia in living reptiles, birds and mammals have many traits in common, but noteworthy for the present analysis are the facts that the basal ganglia is much more cell rich than in amphibians, and receives a much more prominent dopaminergic midbrain input and glutamatergic cortical input than in amphibians (Reiner et al 1998). The stem amniotes must, therefore, have possessed a basal ganglia that shared these features with living amniotes. It seems implausible to think that the stem amniotes could also then have had essentially an amphibian grade of cortical organization, and had such a well-developed subpallial cortical target area (i.e. the basal ganglia).

FIG. 4. (*opposite*) Stage-wise evolution of the pallium in the sauropsid and mammalian lineages, according to the 'temporal neocortex *de novo* hypothesis'. Based on published studies, the amphibian ancestors of stem amniotes are assumed to have lacked a true dorsal cortex or a true V1, both of which are assumed to have evolved in the amphibian to stem amniote transition, based on the wide acceptance of the homology of superior neocortex in mammals and dorsal cortex in reptiles. Ancestral amphibians also must have possessed an extensive laterally situated olfactory pallium (Olf), with a small *Emx1*-negative pallial zone at its lower edge (termed the intermediate zone). This lateral pallial organization must have been inherited by stem amniotes, and the basal ganglia must have been essentially at an amphibian grade of organization due to the poor cortical development presumed by this hypothesis. In the mammalian lineage, therapsids evolved V2 and A1 *de novo* as part of the *de novo* evolution of temporal neocortex. The further changes from the therapsid to the marsupial grade are assumed to consist mainly of thickening and differentiation of the cortical plate, presumably accompanied by the addition of the layer II–IV cell types, and a ventral shift of the olfactory cortex. Finally, the changes in the neocortex from the marsupial to the early placental grade consist mainly of the expansion and further laminar differentiation of the neocortex, and the further ventral shift of olfactory cortex. The intermediate zone of the amphibian lateral pallium evolves into the claustrum, endopiriform and/or basolateral/basomedial amygdaloid regions in the mammalian lineage, according to this hypothesis. In the sauropsid lineage, this hypothesis proposes that the amphibian intermediate pallial zone became the DVR, with V2-like and A1-like areas evolving separately from V2 and A1 of mammalian temporal neocortex. According to this hypothesis, no DVR or proto-DVR was present in stem amniotes, and the DVR bulge evolved gradually in the sauropsid lineage, with at first the DVR possessing a laminar pattern similar to that of the dorsal cortex (*Sphenodon* grade), and the DVR then later losing the laminar organization and taking on the cell group cytoarchitecture evident in turtles, but especially notable in birds. Note that all schematic drawings are of transverse sections through the telencephalon at approximately the level of the anterior commissure. DP, dorsal pallium; S, septum; St, striatum.

Evolutionary transformations — the common origin hypothesis

I shall present here a common origin hypothesis that assumes that the DVR of extant sauropsids and the temporal neocortex of mammals evolved from a proto-DVR condition in which the incipient DVR did not bulge into the ventricle in stem amniotes (Fig. 5). This differs from the view presented by Karten (1969, 1991), in which he appeared to propose a transformation of a DVR into temporal neocortex in the stem amniote–mammal transition (Reiner 1996). From a proto-DVR state, it is here proposed that cortical expansion and translocation of the proto-DVR to a slightly more superficial position occurred in the earliest members of the mammalian lineage (therapsids). The evolutionary changes in the proto-DVR leading to the therapsid grade are assumed to consist mainly in a shift in the final adult positions of the cortical plate, the basal ganglia and the pyriform cortex. The shift in the proto-DVR plate to a cortical position may have been accompanied by an acquisition of *Emx1* expression by the proto-DVR, or conversely it may have expressed *Emx1* even in stem amniotes. Addition of layer II–IV cell types and the shift to an inside-out gradient of neurogenesis must have occurred at some point in the transition from therapsids to mammals (Reiner 1991, 1993). Refinements in

---

FIG. 5. (*opposite*) Proposed evolution of the pallium according to the 'common origin of temporal neocortex and dorsal ventricular ridge (DVR) hypothesis'. For the reasons noted in the legend for Fig. 4, the amphibian ancestors of stem amniotes are assumed to have lacked a true dorsal cortex or a true V1, but have possessed an extensive laterally situated olfactory pallium (Olf), with a small *Emx1*-negative pallial zone at its lower edge (termed the intermediate zone). This hypothesis assumes that a similar lateral pallial organization must have been inherited by stem amniotes, but a V1-containing dorsal cortical zone and a contiguous V2-containing and A1-containing proto-DVR (in which the DVR does not bulge into the ventricle) emerged in stem amniotes as part of the necessary neural processing ability needed for the full emergence to land. As part of this adaptation, the basal ganglia also achieved a level of organization similar to that in living reptiles. These assumptions make it possible to account for the topological similarities in these sensory areas that are evident between modern reptiles and mammals and to account for the seamless transformation of both the stem amniote cortex and proto-DVR into mammalian neocortex. The evolutionary changes in the proto-DVR leading to the therapsid grade are assumed to consist mainly in the shift of the positions of the cortical plate, the basal ganglia and the pyriform cortex. The further changes from the therapsid to the marsupial grade are assumed to consist mainly of thickening and differentiation of the cortical plate, presumably accompanied by the addition of the layer II–IV cell types, and the further ventral shift of the olfactory cortex. Finally, the changes in the neocortex from the marsupial to the early placental grade are assumed to consist mainly of the expansion and further laminar differentiation of the neocortex, and the further ventral shift of the olfactory cortex. The evolutionary changes in the proto-DVR leading to the early sauropsid and *Sphenodon* grades are assumed to consist mainly of enlargement of the DVR and its ingrowth into the ventricle. Finally, the changes in the DVR from the *Sphenodon* to the turtle grade consisted of cytodifferentiation of the DVR and pallial thickening into non-periventricular cell groups. Note that all schematic drawings are of transverse sections through the telencephalon at approximately the level of the anterior commissure. DP, dorsal pallium; S, septum; St, striatum.

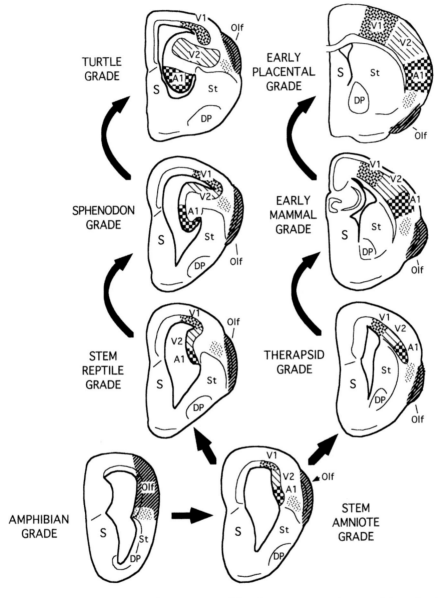

laminar differentiation may have occurred by the monotreme/marsupial grade of evolution, and proliferation of cortical areas occurred separately in the three major mammalian lineages (Northcutt & Kaas 1995). In constructing this hypothesized stem amniote cortical organization, it is assumed that distinct S1, V1, V2 and A1 regions were already present in the dorsal cortex/pallial thickening and proto-DVR of stem amniotes, and that the cells of the proto-DVR and cortex in stem amniotes were both in a periventricular position (Reiner 1993, Butler 1994b). In the sauropsid lineage, the evolutionary changes in the proto-DVR leading to the early sauropsid and *Sphenodon* grades are assumed to consist mainly of enlargement of the DVR and its ingrowth into the ventricle. *Emx1* expression may have been lost in the proto-DVR to rostral DVR transformation, or it may never have been expressed by the proto-DVR. The changes in the DVR from the *Sphenodon* to the turtle grade are assumed to consist of cytodifferentiation of the DVR and pallial thickening into non-periventricular cell groups. The *Emx1*-negative intermediate pallial zone and surrounding tissue found in amphibians may indeed have evolved into the claustrum, endopiriform and/or basolateral amygdala regions in mammals. In the sauropsid lineage, the *Emx1*-negative pallial region may have evolved into a small ill-defined region deep to the pyriform cortex or into parts of the posterior DVR. These various assumptions make it possible to account for the topological similarities in the 'cortical' sensory areas that are evident among modern sauropsids and mammals, to account for the seamless transformation of both stem amniote cortex and proto-DVR into mammalian neocortex by the addition of layer II–IV cell types, and to account for the organization of the DVR in *Sphenodon* compared to that in other reptiles. They also account for the prominent similarities in the V2 and A1 circuits between mammals and sauropsids. Additionally, postulating a proto-DVR in stem amniotes would be consistent with the narrowness of the telencephalon in early members of the mammalian lineage (Hopson 1979). Finally, since this hypothesis posits that stem amniotes possessed a dorsal pallium beyond the amphibian grade, there would be no disparity between basal ganglia and cortex in grade of complexity in the hypothesized stem amniote. I thus believe that the 'common origin of DVR and temporal neocortex hypothesis' currently presents a more parsimonious account of neocortical evolution than does the 'temporal neocortex *de novo* hypothesis'.

Issues for further study

While the evolution of neocortex is likely to remain an area of debate, recent findings and review papers have invigorated progress and discussion of this issue. In terms of the notions presented here, a number of additional lines of data would be useful for evaluating the two hypotheses presented. First, the most compelling evidence for the 'temporal neocortex *de novo* hypothesis' involves the absence of *Emx1* expression from some subcortical pallial regions that border the pyriform cortex and basal ganglia in amphibians, reptiles, birds and mammals. At present it is not possible to infer with certainty from the *Emx1* data that these regions share a common ancestry. It could be that *Emx1* expression has been lost by DVR or gained by the temporal neocortex. Along these lines, it would be desirable to assess the merits of the 'temporal neocortex *de novo* hypothesis' by additional developmental or adult markers. For both hypotheses, the developmental and topological relationship of pyriform (olfactory) cortex to DVR on the one hand, and claustrum, endopiriform region and amygdala on the other needs to be resolved. The *de novo* hypothesis typically posits that these subcortical pallial regions arise at the ventral edge of the proliferative zone giving rise to olfactory cortex, while the 'common origin hypothesis' posits that olfactory cortex arises at the ventral edge of the proliferative zone giving rise to DVR and temporal neocortex. Finally, detailed studies of the mesencephalic and diencephalic components of the visual and auditory pathways to the telencephalon are needed for amphibians. These circuits exist in amphibians, but the thalamic projections end in the striatum for the V2-type and A1-type pathways (Northcutt & Kicliter 1980, Bruce & Neary 1995). It is unclear if these are intralaminar type circuits, or if these are homologous to the V2 and A1 circuits in reptiles and birds, with the thalamic fibres of these amphibian circuits coming to invade the pallium during the amphibian to stem amniote transition. It is important to examine the mesencephalic and diencephalic components of these circuits to ascertain how similar they truly are to the mesencephalic and diencephalic components of the V2 and A1 circuits in reptiles and birds. If those in amphibians are highly similar to those in reptiles and amphibians, the possibility would be raised that the similarities between the V2 and A1 circuits of mammals to those in sauropsids is an instance of parallel evolution channelled by the already existing state of development of these circuits in ancestral amphibians and stem amniotes.

Acknowledgements

I thank Harvey J. Karten, Steven E. Brauth, R. Glenn Northcutt, Loreta Medina and Luis Puelles for their provocative discussions with me on the topic of forebrain evolution over the years. In particular the proto-DVR hypothesis is an outgrowth of discussions that Steve Brauth and I had a number of years ago. I also thank R. Glenn Northcutt for allowing me to use some of his unpublished *Sphenodon* data in this paper. Finally, I would like to thank Luis Puelles for providing me with a pre-print of a paper reporting his intriguing recent homeobox data. The research from my laboratory presented here has been supported by NS-19620, NS-28721 and EY-05298.

References

Allman J 1990 Evolution of neocortex. In: Jones EG, Peters A (eds) Cerebral cortex, vol 8a: Comparative structure and evolution of cerebral cortex, part 1. Plenum Press, New York, p 269–283

Arbib MA, Erdi P, Szentagothai J 1998 Neural organization: structure, function and dynamics. MIT Press, Cambridge, MA

Arimatsu Y, Kojima M, lshida M 1999 Area- and lamina-specific organization of a neuronal subpopulation defined by expression of latexin in the rat cerebral cortex. Neuroscience 88:93–105

Balaban CD, Ulinski PS 1981 Organization of thalamic afferents to anterior dorsal ventricular ridge in turtles. I. Projection of thalamic nuclei. J Comp Neurol 200:95–129

Beck P, Pospichal M, Kaas JH 1996 Topography, architecture, and connections of somatosensory cortex in opossums: evidence for five somatosensory areas. J Comp Neurol 366:109–133

Brauth SE, Reiner A 1991 Calcitonin-gene related peptide is an evolutionarily conserved marker within the amniote thalamo-telencephalic auditory pathway. J Comp Neurol 313:279–239

Bruce LL, Butler AB 1984a Telencephalic connections in lizards. I. Projections to cortex. J Comp Neurol 229:585–601

Bruce LL, Butler AB 1984b Telencephalic connections in lizards. II. Projections to anterior dorsal ventricular ridge. J Comp Neurol 229:602–615

Bruce LL, Neary TJ 1995 The limbic system of tetrapods: a comparative analysis of cortical and amygdalar populations. Brain Behav Evol 46:224–234

Butler AB 1994a The evolution of the dorsal thalamus of jawed vertebrates, including mammals: cladistic analysis and a new hypothesis. Brain Res Brain Res Rev 19:29–65

Butler AB 1994b The evolution of the dorsal pallium in the telencephalon of amniotes: cladistic analysis and a new hypothesis. Brain Res Brain Res Rev 19:66–101

Coleman J, Clerici WJ 1981 Organization of thalamic projections to visual cortex in opossum. Brain Behav Evol 18:41–59

Diamond IT, Hall WC 1969 Evolution of neocortex. Science 164:251–262

Durward A 1930 The cell masses in the forebrain of *Sphenodon punctatum*. J Anat 65:8–44

Ebner FF 1976 The forebrain of reptiles and mammals. In: Masterton RB, Bitterman ME, Campbell CBG, Hotton N (eds) Evolution of brain and behavior in vertebrates. Lawrence Erlbaum Associates, Hillsdale, NJ, p 147–167

Elliot-Smith G 1919 A preliminary note on the morphology of the corpus striatum and the origin of the neopallium. J Anat 53:271–291

Fernández A, Pieau C, Repérant J, Boncinelli E, Wassef M 1998 Expression of the *Emx-1* and *Dlx-1* homeobox genes define three molecularly distinct domains in the telencephalon of mouse, chick, turtle and frog embryos: implications for the evolution of telencephalic subdivisions in amniotes. Development 125:2099–2111

Hall JA, Foster RE, Ebner FF, Hall WC 1977 Visual cortex in a reptile, the turtle (*Pseudemys scripta* and *Chrysemys picta*). Brain Res 130:197–216

Hodos W 1982 Some perspectives on the evolution of intelligence and the brain. In: Griffin DR (ed) Animal mind–human mind. Springer-Verlag, New York (Dahlem Konferenzen) p 34–55

Holmgren N 1925 Points of view concerning forebrain morphology in higher vertebrates. Acta Zool Stockh 6:413–477

Hopson JA 1979 Paleoneurology. In: Gans C, Northcutt RG, Ulinski P (eds) Biology of the Reptilia, vol 9: Neurology A. Academic Press, New York, p 39–146

Jerison HJ 1985 Animal intelligence as encephalization. Phil Trans R Soc Lond B Biol Sci 308:21–35

Johnston JB 1915 The cell masses in the forebrain of the turtle, *Cistudo carolina*. J Comp Neurol 25:393–468

Jones EG 1981 Anatomy of cerebral cortex: columnar input–output organization. In: Schmitt FO, Worden F, Adelman G, Dennis SG (eds) The organization of the cerebral cortex. MIT Press, Cambridge, MA, p 199–235

Kaas JH 1980 A comparative survey of visual cortex organization in mammals. In: Ebbesson SOE (ed) Comparative neurology of the telencephalon. Plenum Press, New York, p 483–502

Källén B 1951 The nuclear development in the mammalian forebrain with special regard to the subpallium. K Fysiogr Sallsk Lund Forh 62:1–43

Källén B 1962 Embryogenesis of brain nuclei in the chick telencephalon. Ergeb Anat Entwicklungsgesch 36:62–82

Karten HJ 1969 The organization of the avian telencephalon and some speculations on the phylogeny of the amniote telencephalon. Ann NY Acad Sci 167:164–179

Karten HJ 1991 Homology and evolutionary origins of the 'neocortex'. Brain Behav Evol 38:264–272

Luksch H, Cox K, Karten HJ 1998 Bottlebrush endings and large dendritic fields: motion detecting neurons in the tectofugal pathway. J Comp Neurol 396:399–414

Macphail EM 1982 Brain and intelligence in vertebrates. Clarendon Press, Oxford

Major DE, Luksch H, Karten HJ 1998 Multiple subtypes of neurons in the lower stratum griseum superficiale of the ground squirrel superior colliculus revealed by in vitro cell filling. Soc Neurosci Abstr 24:1631

Marín-Padilla M 1998 Cajal-Retzius cells and the development of neocortex. Trends Neurosci 21:64–71

Nauta WJH, Karten HJ 1970 A general profile of the vertebrate brain, with sidelights on the ancestry of cerebral cortex. In: Schmitt FO (ed) The neurosciences: second study program. Rockefeller University Press, New York, p 7–26

Northcutt RG 1981 Evolution of the telencephalon in nonmammals. Annu Rev Neurosci 4:301–350

Northcutt RG, Kaas JH 1995 The emergence and evolution of mammalian neocortex. Trends Neurosci 18:373–379

Northcutt RG, Kicliter E 1980 Organization of the amphibian telencephalon. In: Ebbesson SOE (ed) Comparative neurology of the telencephalon. Plenum Press, New York, p 203–255

Orrego F 1961 The reptilian forebrain. I. The olfactory pathways and cortical areas in the turtle. Arch Ital Biol 99:425–445

Pobirsky N, Molnár Z, Blakemore C, Krubitzer L 1998 The organization of somatosensory cortex in the western European hedgehog (Erinaceus europaeus). Soc Neurosci Abstr 24:1125

Pritz MB 1980 Parallels in the organization of auditory and visual systems in crocodiles. In: Ebbesson SOE (ed) Comparative neurology of the telencephalon. Plenum Press, New York, p 331–342

Puelles L, Rubinstein JLR 1995 Expression patterns of homeobox and other putative regulatory genes in the embryonic mouse forebrain suggest a neuromeric organization. Trends Neurosci 16:472–479

Puelles L, Kuwana E, Puelles E et al 2000 Pallial and subpallial derivative in the embryonic chick and mouse telencephalon, traced by the expression of the genes Dlx-2, Emx-1, Nkx-2.1, Pax-6 and Tbr-1. submitted

Reiner A 1991 A comparison of the neurotransmitter-specific and neuropeptide-specific neuronal cell types present in turtle cortex to those present in mammalian isocortex: implications for the evolution of isocortex. Brain Behav Evol 38:53–91

Reiner A 1993 Neurotransmitter organization and connections of turtle cortex: implications for the evolution of mammalian isocortex. Comp Biochem Physiol A Comp Physiol 104:735–748

Reiner A 1994 Laminar distribution of the cells of origin of the ascending and descending tectofugal pathways in turtles: implications for the evolution of tectal lamination. Brain Behav Evol 43:254–292

Reiner A 1996 Levels of organization and the evolution of isocortex: homology, nonhomology or parallel homoplasy. Trends Neurosci 19:89–91

Reiner A, Brauth SE, Karten HJ 1984 Evolution of the amniote basal ganglia. Trends Neurosci 7:320–325

Reiner A, Medina L, Veenman CL 1998 Structural and functional evolution of the basal ganglia in vertebrates. Brain Res Rev 28:235–285

Rowe M 1990 Organization of the cerebral cortex in monotremes and marsupials. In: Jones EG, Peters A (eds) Cerebral cortex, vol 8b: Comparative structure and evolution of cerebral cortex. Plenum Press, New York, p 263–334

Striedter GF 1997 The telencephalon of tetrapods in evolution. Brain Behav Evol 49:179–213

Swanson LW, Petrovich GD 1998 What is the amygdala? Trends Neurosci 21:323–330

Ulinski PS 1988 Functional architecture of turtle dorsal cortex. In: Schwerdtfeger WK, Smeets WJAJ (eds) The forebrain of reptiles. Current concepts of structure and function. Karger, Basel, p 151–161

Ulinski PS 1990 Nodal events in forebrain evolution. Neth J Zool 40:215–240

Veenman CL, Medina L, Reiner A 1997 Avian homologues of the mammalian intralaminar, mediodorsal and midline thalamic nuclei: immunohistochemical and hodological evidence. Brain Behav Evol 49:78–98

DISCUSSION

Pettigrew: Could you tell us why you call this temporal visual area V2, because I think of V2 as being a mirror-image representation of V1, and not isolated far away in the temporal region as your V2 is.

Reiner: By V2 I mean the tecto-thalamocortical pathway.

Karten: This may be a poor choice of term because V2 has a different meaning for the vast majority of people in this field. The area you're talking about probably corresponds to temporal visual cortical areas.

Kaas: I don't agree that it's a poor choice because V2 is an area common to almost all mammals, so it must have been present in reptiles.

Karten: Recently, I spend much of my time working on this, and what I would argue is that the tectofugal pathway that Anton Reiner refers to as V2 is the targeted pathway coming out of the caudal or inferior pulvinar in squirrels and going on to the areas you and Bill Hall originally called TP in the squirrel. We believe this is equivalent to the inferotemporal cortex in primates, but it is not equivalent to the region we call V2 in primates, because V2 is mainly area 18. As a general notion, calling the two major visual pathways V1 and V2 is sensible, but because of the encumbrances of other terminologies we would probably be wise in naming them differently.

Kaas: But I am suggesting that this area really is V2. There is a paucity of evidence, so we shouldn't rule out the other possibility that it is a temporal visual area.

Pettigrew: I like that because there is short latency, early developing visual input in the temporal lobe that could be equivalent to area MT (middle temporal visual area). There is tremendous controversy on this, but I subscribe to the view that primate MT is the tectofugal visual destination.

Puelles: We should first find out whether or not such a telencephalic area in sauropsids is homologous to mammalian cortex. If we find that it is, we can then discuss which area it represents.

Pettigrew: I would like to add that Marcello Rosa has found that MT in primates is surrounded by a crescent which is the mirror-symmetrical in the same way as V2 is a mirror-symmetrical crescent around V1. This strongly supports the idea that MT is a separate system.

Molnár: There are no disagreements about the numerous similarities in the physiological properties, receptive field characteristics of single units and the functional maps of Wulst of birds, dorsal ventricular ridge (DVR) of reptiles and extrastriate cortex in mammals. All the recording studies, which were done in the pigeon and more recently in the iguana (Manger et al 1997), show clearly that it has similar properties as a secondary visual area. The disagreements concern the origin of the cells constituting these structures.

Krubitzer: We have demonstrated that in the iguana there are several representations of the visual field in the DVR, and neurons here respond vigorously. There are reversals in receptor field progression across the borders of fields. Also, we see receptive field configurations that are similar to those seen in the visual cortex of mammals. However, the neurons in the iguana cortex respond poorly to visual stimulation. The receptive fields of neurons in the DVR are similar to those of neurons in V1 rather than V2. In mammals, the response properties fall off dramatically as you exit V1, so I'm not quite certain if Anton has positioned V1 correctly relative to V2. I would like to ask Anton Reiner to address this because his hypothesis depends upon the geographic relationship between these fields in mammals and in turtles.

Reiner: In order to comment on this properly I would have to see where the V1 is located in your preparations. We do know that the V1 differs in iguanas and turtles. In turtles, V1 has a slight subcortical extension. It is partly at the surface and partly subcortical. However, this extension is magnified in iguanas, so that the whole of V1 has a subcortical localization and is translocated to the DVR.

Molnár: Tony, you mentioned that the temporal cortex *de novo* hypothesis and the common origin hypothesis both state that the dorsal cortex is of reptiles and dorsal cortex of mammals have common cells. However, it is not at all clear in the literature exactly which cells are common (Marín-Padilla 1971, Reiner 1993). Could you comment on the controversy concerning the extent to which pre-existing structures contribute to the mammalian isocortex.

Reiner: The differences in opinion centre on whether or not the cells in the dorsal cortex are entirely preplate cells and there are no other cell types present. I cannot resolve this issue because we need some fine tools to distinguish cell types in the dorsal cortex more carefully than has been done in the past. My own histochemical work suggests that there are more types of cells in the dorsal cortex than are found

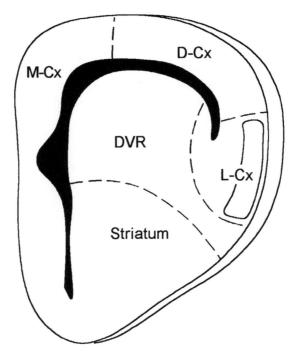

FIG. 1.   Schematic cross-section through the reptilian telencephalon, illustrating the relative topology of the dorsal ventricular ridge (DVR), lateral cortex (L-Cx) and dorsal cortex (D-Cx), as indicated by radial glial processes (C. M. Yanes & L. Puelles, unpublished observations 1999) and observations in *Sphenodon* (Fig. 2 shown by Reiner). Note that the lateral cortex relates to a radial field intercalated between the DVR and dorsal cortex. This arrangement seems to imply that any postulated contribution of DVR-originated cells to the developing dorsal cortex necessarily must proceed through tangential migration underneath the lateral cortex. M-Cx, medial cortex.

in the preplate. Also, the recent turtle literature suggests that there is some formation of cortical plate (Goffinet et al 1986, Marín-Padilla 1998). The turtle dorsal cortex is equivalent only to layers V and VI in the mammalian neocortex, and the major evolutionary change that took place between turtle dorsal cortex and mammalian neocortex is the appearance of layer II–IV cell types and of course the inside-out migration pattern.

Puelles: I would like to say a few words about the topology of the DVR versus the lateral and dorsal pallium and the hypothesis presented by Anton Reiner, which I believe has some difficult implications, also shared by Harvey Karten's ideas. Figure 1 shows the DVR, the pallium, the subpallium and the lateral cortex. One topological problem with Anton Reiner's hypothesis is that he apparently assumes there is no relationship between the lateral cortex at the brain surface and the corresponding ventricular zone. Therefore, it is possible for him to distort topologically the telencephalic wall in his drawings and assign cellular material evolving first within the DVR (in reptiles) to a position overlying that area in mammals (temporal isocortex). I'd like to point out that we first need to use available data showing that the ventricular ridge relates radially to a narrow part of the brain surface intercalated between the subpallium and the lateral cortex, which is the topological position of the lateral olfactory tract. In contrast, the overlying subpial lateral cortex relates radially to the ventricular zone around the lateral ventricular angle, i.e. this radial complex overlies topologically the DVR and separates it wholly from the dorsal pallium, the site where the isocortex is supposed to emerge. This fundamental topological relationship between subpallium, DVR, lateral cortex and isocortex is deformed morphogenetically in many reptiles, so there is a lot of bending of the radial dimension, and this is exemplified by the DVR bulge itself. This reptilian pallial pattern nevertheless has to be defined as ancestral for the mammalian pallium and the evolution of the cortex. We have been looking at the radial glia in this territory (C. M. Yanes & L. Puelles, unpublished observations 1999) and the data clearly corroborate the above statements on relative topological positions. This explains what you saw in Anton Reiner's slide of the *Sphenodon* telencephalon, i.e. where there is scarcely any migration of pallial neurons to subpial strata. The *Sphenodon* unmigrated lateral cortex clearly separates the dorsal cortex from the DVR. This pattern is to be found in all reptiles and also in birds. It forces the conclusion that, in order for any DVR-originated cell population to incorporate into, or build, any part of the mammalian temporal cortex, tangential cell migration must occur during cortical development to bring these cells under the lateral cortex, and past the rhinal sulcus, into the temporal cortex. I do not see how this topological reasoning can be escaped. The problem with Karten's and Reiner's hypothesis is that there is no evidence for this migration in any mammal (even though other tangential migrations have been discovered; Anderson et al 1997). Reiner's drawings, disregarding this point, tend to give the impression that cell movement from DVR to the temporal cortex can be a straightforward radial migration.

Herrup: If you moved either anteriorly or posteriorly from that section, is there a region where migration such as you describe is less difficult because the cells can move around it rather than through it?

Butler: I agree that the topography problem concerning the olfactory cortex has to be resolved, and I can't see a way of doing this yet. However, I would like to raise a couple of points concerning the dorsal thalamus, because here I can make a case for conservative evolution of this part of the forebrain across vertebrates, including all amniotes, amphibians and fish. Therefore, if there were major transformations in the telencephalon, we have to account for them, especially in regard to the idea that the DVR could be similar to the amygdala or the claustrum, and possible major differences in sensory connections between mammals and other amniotes.

About five or six years ago, Anton Reiner and I independently came up with the idea he talked about today of evolution of the DVR in reptiles versus the lateral neocortex in mammals. Both of these structures are specializations; neither were present in the common ancestor. However, the common ancestor had sensory pathways for the visual and auditory systems going to the striatum and to part of the dorsolateral pallium similar to the LP/pulvinar and medial geniculate pathways in mammals and the rotundus and ovoidalis pathways in birds and reptiles. Likewise, a dorsal lateral geniculate type of visual pathway was present.

There are two recognizable areas of the anamniote dorsal thalamus (Braford & Northcutt 1983, Neary & Northcutt 1983, Butler & Northcutt 1992). There is a visual–somatosensory area in the rostral thalamus, nucleus anterior, which receives visual input directly from the retina. This area does not receive its major input from the midbrain roof. Caudal to this area, there are two nuclei, called DP (dorsal posterior) and CP (central posterior) in fish. DP gets visual input from the superior colliculus homologue, whereas CP gets auditory input from the inferior colliculus homologue. I call DP and CP the collothalamus, and I call nucleus anterior the lemnothalamus, 'lemno' referring to lemniscal (i.e. direct) input that does not stop off in the midbrain along the way. These two divisions are also apparent in mammals, reptiles and birds.

In 1942 Rose published a paper on the rabbit diencephalon. At certain stages of development, he saw what he called 'pronuclei' that gave rise to other pronuclei or to definitive nuclei in the adult. He saw a dorsal pronucleus and a medial geniculate pronucleus (Rose 1942). The dorsal pronucleus gives rise to the lateral posterior nucleus and the posterior nuclear group, and the medial geniculate pronucleus, which develops at about the time of birth, gives rise to the medial geniculate nucleus. These two pronuclei together form what I call the collothalamus. We are now analysing this in rats, and we found that one rostral pronuclear mass gives rise to the dorsal lateral geniculate nucleus and to the rest of the lemnothalamic nuclei, including the entire ventral nuclear group. Regarding the two pronuclei that form the collothalamus, this situation is virtually the same as in anamniotes, where DP—similar to the nuclei that derive from the dorsal pronucleus—gets visual and somatosensory input from the superior colliculus homologue and CP—just like the medial geniculate—gets auditory tectal input.

We looked at the rat at embryonic day (E) 17, when only a large mass of a single lemnothalamic pronucleus is present. At E19, a small rostral pole of a separate cell mass (LP) appears that can be followed caudally and that is derived from the dorsal pronucleus. The dorsal pronucleus gives rise to the lateral posterior nucleus and the posterior nuclear group. The medial geniculate nucleus forms at this caudal level more ventrally and laterally.

A collothalamic pronuclear mass is also present in birds, and in this case we call it RTO. It will split medially into the auditory nucleus (nucleus ovoidalis) and laterally into nuclei rotundus and triangularis. These are separate and distinct from the more rostral, developing lemnothalamus. In birds the collothalamus dominates, whereas in mammals the lemnothalamus dominates. The point is that it is possible to compare the thalamus in reptiles and birds versus mammals; they have similar nuclear groups, they develop in the same way and they have the same two major divisions. Any theories of telencephalic evolution are going to have to account for the pattern of projections that they send up to the telencephalon.

Kaas: The crux of these arguments seems to be the issue of connections, and I wonder if this can be addressed by proposing that auditory projections were originally to both subcortical structures and to dorsal cortex, and that one connection then became more emphasized.

Karten: We often tend to concentrate on issues pertaining to the visual system because there is so much work being done on the visual system, but we don't tend to resolve issues relating to the evolutionary framework. In contrast, the auditory system seems to be much more limited in its diversity, but if we compare the auditory system in birds and mammals, the similarities in organization and coding are striking. I wonder whether we should be focusing on the auditory system because there's far less controversy in this field than in the visual system.

Levitt: Isn't there a direct thalamic projection to the amygdala and a direct thalamic projection to auditory cortex?

Karten: There is some degree of controversy about the nature of the projections of the auditory system to the amygdala, in terms of where they end, how they are organized, and whether they come from the ventral nucleus of the medial geniculate rather than from the medial portion of the medial geniculate.

Levitt: But in the thalamic zone of these lower vertebrates, how can you differentiate between overlapping projections that may later separate into cortical and amygdala representations?

Karten: In the auditory system there is no evidence for anything that even vaguely resembles a tonotopic organization of auditory input to the amygdala, whereas there is evidence for this in field L. There is no evidence of a tonotopic organization or discrete auditory input into the claustrum, whereas in field L and the mammalian auditory cortex there is.

Levitt: LeDoux and colleagues have shown that in a model of fear conditioning in rodents, it is possible to record from auditory thalamus to lateral amygdala nucleus and auditory cortex (Rogan & LeDoux 1996). They demonstrate a crude tonotopic map in the lateral amygdala. It is possible to shift this map, based on conditioning in which a conditioning stimulus, such as a specific tone, is paired with an aversive stimulus, such as a foot shock. Neurons show the ability to respond in a similar way in terms of shifting their best frequency to that of the conditioning stimulus in the thalamus, cortex and even in the amygdala. If it is the case that we have a unique situation in which there are direct projections from auditory thalamus to amygdala, if indeed there is some degree of tonotopy, and if the best frequency can be shifted here and in the cortex, how is it possible to use connectivity to define these regions?

Karten: What's seen in field L is entirely different. Field L has extremely precise tonotopy. It is sharply tuned, and the responses of the cells resemble those of layer IV in the cortex. Therefore, it is a different circumstance. There is spectral response differentiation shown in the amygdala, but it is not truly an auditory map in the same sense.

Puelles: Is it possible that the different demands on the function cause evolution of a different structure with only roughly similar characteristics?

Karten: That is possible, but the projections come from a different subdivision of the medial geniculate.

References

Anderson SA, Eisenstat D, Shi L, Rubenstein JLR 1997 Interneuron migration from basal forebrain: dependence on *Dlx* genes. Science 278:474–476

Braford MR Jr, Northcutt RG 1983 Organization of the diencephalon and pretectum of the ray-finned fishes. In: Davis RE, Northcutt RG (eds) Fish neurobiology, vol 2. University of Michigan Press, Ann Arbor, MI, p 117–164

Butler AB, Northcutt RG 1992 Retinal projections in the bowfin, *Amia calva*: cytoarchitectonic and experimental analysis. Brain Behav Evol 39:169–194

Goffinet AM, Daumierie CH, Langerwerf B, Pieau C 1986 Neurogenesis in reptilian cortical structures: ^{3}H-thymidine autoradiographic analysis. J Comp Neurol 243:106–116

Marín-Padilla M 1971 Early prenatal ontogenesis of the cerebral cortex (neocortex) of the cat (*Felis domestica*). A Golgi study. I. The primordial neocortical organization. Z Anat Entwicklungsgesch 134:117–145

Marín-Padilla M 1998 Cajal-Retzius cells and the development of the neocortex. Trends Neurosci 21:64–71

Manger P, Molnár Z, Slutsky D, Krubitzer L 1997 Subdivisions of visually responsive regions of the dorsal ventricular ridge of the iguana (*Iguana iguana*). Soc Neurosci Abstr 23:1032

Neary TJ, Northcutt RG 1983 Nuclear organization of the bullfrog diencephalon. J Comp Neurol 213:262–278

Reiner A 1993 Neurotransmitter organization and connections of turtle cortex: implications for the evolution of mammalian isocortex. Comp Biochem Physiol A Comp Physiol 104:735–748

Rogan MT, LeDoux JE 1996 Emotion: systems, cells, synaptic plasticity. Cell 85:469–475

Rose JE 1942 The ontogenetic development of the rabbit's diencephalon. J Comp Neurol 77:61–129

General discussion II

Amniote evolution

Evans: I would like to say a few words about how palaeontologists see the relationships between the modern groups. If someone had asked me to talk about the relationships between birds and mammals 40 years ago, I might have drawn a diagram showing the two groups coming out of reptiles (Fig. 1). However, over the last decade, we have produced a more precise picture (Fig. 2). Lizards, tuataras (*Sphenodon* and its relatives), crocodiles and birds are grouped together, and mammals are down at the bottom. This is the palaeontological consensus, i.e. the morphological consensus, and it is also the molecular consensus (e.g. Lee 1993, Reisz 1997). The problem group has been turtles, with much debate as to their relationships. Both the morphological and the molecular consensus would place turtles in the clade crownwards of mammals. Personally, I would place them as shown in Fig. 2, and this is the consensus view. This has a number of implications. First, Reptilia is a distinct clade, and the concept of a stem reptile giving rise to mammals is outdated; stem reptiles gave rise to other reptiles, including birds. If we are talking about the group ancestral to reptiles and mammals, we should be talking about stem amniotes and not stem reptiles. You may think this is just semantics, but it is important to avoid confusion and possible mistakes. Also, if you are trying to work out what the stem mammalian (synapsid) condition was like, you have to bracket it between the living groups available, i.e. amphibians, the most primitive living mammals (such as monotremes) and the most basal reptiles.

I would query the suggestion that the distance between mammals and amphibians is too great to make comparison useful because the palaeontological evidence suggests a considerable pulse of tetrapod evolution within quite a short space of time. The amphibian and amniote lineages separated from one another around 340–350 million years ago, but the last common ancestor of mammals and reptiles lived about 320 million years ago, so the difference between the two separation points is not that great. The ancestors of diapsids and turtles separated at least 300–310 million years ago. The lineages leading to birds and crocodiles separated around 235 million years ago, and the lines leading to lizards and snakes on the one hand and tuataras on the other split at least 220 million years

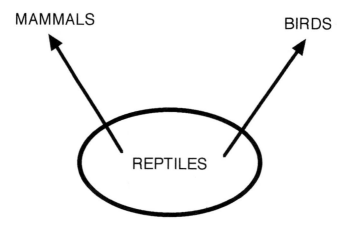

FIG. 1.   Schematic diagram showing general consensus of amniote relationships around 40 years ago.

ago. (These dates are based on the earliest known occurrence of fossils representing the descendant lineages.)

Recently Rieppel & deBraga (1996, deBraga & Rieppel 1997) suggested that turtles might belong within Diapsida, close to lizards and tuataras, but their data matrix has subsequently been corrected by other authors (Lee 1997), and when the analysis is rerun (Lee 1997, Wilkinson et al 1997), the turtles come out in the traditional position, i.e. on the reptilian stem. However, there have been a number of recent molecular studies, e.g. Platz & Conlon (1997), Zardoya & Meyer (1998), Hedges & Poling (1999), which suggest that turtles should be placed in a different position, i.e. with crocodiles and birds. Why turtles are ending up in this position is not clear. Eernisse & Kluge (1993) did a total evidence analysis, i.e. they put the molecular and morphological data together to find the best-fit tree, and found that the combined evidence produced the traditional tree, i.e. turtles outside Diapsida. Hedges & Poling (1999) also suggested that *Sphenodon* should be close to crocodiles and birds, which I find totally untenable. I have been working on the fossil evidence of this group for a long time, and it is not possible that they belong in Hedges' position. *Sphenodon* is not a primitive animal. It has some primitive features, but shares many advanced characters with lizards.

Reiner: *Sphenodon* certainly shares the lizard characteristic of having a flipped over cerebellum, which is not found in any other reptile. Another striking aspect of *Sphenodon* is that they have a primitive dorsal ventricular ridge (DVR). Have you any thoughts about how they could be so far up in reptile phylogeny, and yet have this unique telencephalic formation?

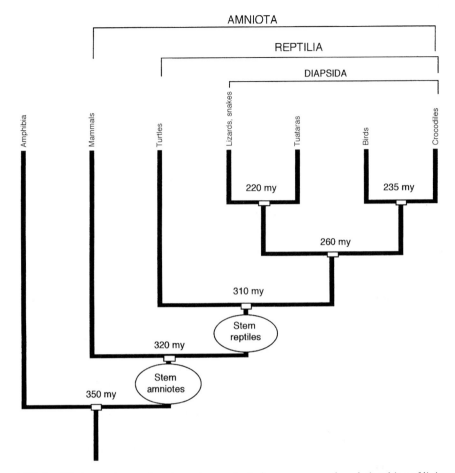

FIG. 2.   Phylogenetic tree showing palaeontological consensus on the relationships of living tetrapod groups. The dates at nodes are in millions of years (my) and are based on the earliest known record of one or both of the lineages diverging from that node. These dates may move back in time as new fossils are found; they are unlikely to move in the opposite direction.

Evans: I can't answer that. It is possible that tuataras retained the primitive condition in this feature. I would like to see what is happening in amphibians, because this issue has not been discussed.

Karten: One problem is that the amphibian brain is difficult to work with. The neurons often tend to be near the ventricle, they don't have large numbers of differentiated populations and they don't have a DVR to speak of. Our concern is, even though they are the only representatives, to what degree are they valid representatives of what occurred at that grey point 350 million years ago? Are they so specialized that they are not good models?

Puelles: They do express genes that are characteristic of the DVR.

Karten: Do they express *Tbr*?

Papalopulu: They express a *Tbr*-related gene, and they also express *Emx1*.

Karten: Can we consider those genes to be comparable genes? What does it mean when we talk about *Emx* in birds, mammals and amphibians? They are not really the 'same' because they are not in the same genome and they do not have identical sequences.

Papalopulu: They are not the same genes, but they may have a common ancestral gene.

Herrup: We have to be careful about dismissing homologies. It is possible to insert the *Drosophila engrailed* gene into a mouse and it works. This has to be homology.

Levitt: There is homology in gene structure and homology in patterns of expression. There are plenty of examples of genes that are highly conserved in species, and yet their expression patterns change in different species.

Rubenstein: I am surprised at how minor the differences are in the expression patterns of all the homologous genes we've looked at during similar stages of development in different species.

Levitt: That may be true for transcription factors.

Puelles: Oscar Marín and collaborators published a paper on the frog basal ganglia (Marín et al 1998) in which they showed amygdala-like expression properties of some neuropeptides in a small ventral pallial domain overlying the striatum. This domain also happens to lack expression of *Emx1* in the frog brain, as occurs in the epistriatal DVR of the turtle and chick, as well as in a thin area adjacent to the mouse striatum, which may correlate with claustroamygdaloid primordia (Fernández et al 1998, L. Puelles, E. Puelles & G. Gonzalez, unpublished data 1999). These peptide and gene expression patterns agree with the claustroamygdaloid molecular properties of the small stretch of pallium found just over the striatum in frogs. This area had been largely disregarded in previous comparative studies (but see Northcutt 1995).

Levitt: But there is a high degree of structural conservation, both at the gene and protein levels, in axon guidance molecules, yet there are pronounced differences in the expression patterns across species. There are many examples of this, e.g. cadherins, Ig super-family members and integrins.

References

deBraga M, Rieppel O 1997 Reptile phylogeny and the interrelationships of turtles. Zool J Linn Soc 120:281–354

Eernisse DJ, Kluge AG 1993 Taxonomic congruence versus total evidence, and amniote phylogeny inferred from fossils, molecules, and morphology. Mol Biol Evol 10:1170–1195

Fernández AS, Pieau C, Repérant J, Boncinelli E, Wassef M 1988 Expression of the *Emx-1* and *Dlx-1* homeobox genes define three molecularly distinct domains in the telencephalon of mouse, chick, turtle and frog embryos: implications for the evolution of telencephalic subdivisions in amniotes. Development 125:2099–2111

Hedges SB, Poling LL 1999 A molecular phylogeny of reptiles. Science 283:998–1001

Lee MSY 1993 The origin of the turtle body plan: bridging a famous morphological gap. Science 261:1716–1720

Lee MSY 1997 Reptile relationships turn turtle. Nature 389:245–246

Marín O, Smeets WJ, González A 1998 Basal ganglia organization in amphibians: chemoarchitecture. J Comp Neurol 392:285–312

Northcutt RG 1995 The forebrain of gnathostomes: in search of a morphotype. Brain Behav Evol 46:275–318

Platz JE, Conlon JM 1997 . . . and turn back again. Nature 389:246

Reisz RR 1997 The origin and early evolutionary history of amniotes. Trends Ecol Evol 12: 218–222

Rieppel O, deBraga M 1996 Turtles as diapsid reptiles. Nature 384:453–455

Wilkinson M, Thorley J, Benton MJ 1997 Uncertain turtle relationships. Nature 387:466

Zardoya R, Meyer A 1998 Complete mitochondrial genome suggests diapsid affinities of turtles. Proc Natl Acad Sci USA 95:14226–14231

Evolution of cortical lamination: the reelin/Dab1 pathway

Isabella Bar and André M. Goffinet[1]

Neurobiology Unit, University of Namur Medical School, 61 rue de Bruxelles, B5000 Namur, Belgium

Abstract. The mammalian cortical plate is characterized by its radial organization and its inside-outside developmental gradient. Observations on reelin and Dab1-deficient mice show that reelin and Dab1 are both required to develop radial cortical organization and a normal maturation gradient. In the reptilian cortex, radial organization varies among species; it is the most rudimentary in turtles and the most elaborate in lizards, and can be described as intermediate in other species such as crocodilians and *Sphenodon*. On the other hand, the gradient of corticogenesis is directed from outside to inside in all reptiles studied, as well as in mice that are deficient in reelin, Dab1, as well as cyclin-dependent kinase 5 (Cdk5) and p35. All reptiles, even turtles, have reelin-expressing cells in the embryonic marginal zone. Mammals are characterized by a drastic increase in the number of reelin-positive cells (Cajal-Retzius cells) as well as by an amplification of reelin expression per cell. In lizards, the pattern of reelin expression is different, as reelin-expressing cells are also present below the cortical plate. In all mammalian and reptilian species, Dab1 is expressed in cortical plate cells. These data suggest that the reelin/Dab1 pathway was a driver of cortical evolution on the synapsid lineage and that similarities in radial cortical organization between squamates and mammals result from evolutionary convergence.

2000 Evolutionary developmental biology of the cerebral cortex. Wiley, Chichester (Novartis Foundation Symposium 228) p 114–128

The cloning of genes implicated in the control of mammalian cortical development raises issues relevant to cortical evolution. More specifically, the key role played by reelin and Dab1 in the laminar, orderly development of the mammalian cortex suggests that these genes acted as drivers of cortical evolution in the synapsid lineage. Although living reptiles do not bear any direct ancestral relationship to mammals, it is reasonable to assume that comparisons of cortical development in

---

[1]This chapter was presented at the symposium by André M. Goffinet, to whom correspondence should be addressed.

mammals and living reptiles may shed some light on the processes involved in cortical evolution.

In all mammals studied thus far, the cerebral cortex develops according to a common sequence (Caviness 1982, Caviness & Rakic 1978, Lambert de Rouvroit & Goffinet 1998a). It begins with the appearance of an horizontal network, named the preplate, formed of Cajal-Retzius cells, subplate neurons and other less-characterized neuronal types (Sheppard & Pearlman 1997, Meyer et al 1998). The preplate is divided into the external, marginal zone (future layer I) and the internal subplate (future layer VIb) by the migration of the elements of the cortical plate. The cortical plate is the precursor of cortical layers II to VI. The mammalian cortical plate is characterized by two key features. First, its constituent neurons assume an elaborate radial organization. Second, its maturation proceeds from inside to outside; that is, late-generated neurons migrate past previously deposited layers and settle at progressively more superficial levels. These two characteristics are not independent, as mutations in either the reelin or Dab1 genes lead to anomalies of the radial organization of cortical plate neurons as well as to inversion of its maturation gradient. In the cortical plate of reelin- or Dab1-deficient mice, older neurons settle in subpial position and younger cells form more internal layers (Caviness 1982, Gonzales et al 1997, Lambert de Rouvroit & Goffinet 1998a). The fact that reelin or Dab1 mutations disturb both radial organization and neurogenetic gradients shows that these two developmental events are regulated by common mechanisms.

Several years ago, we initiated comparative studies of cortical histogenesis in reptiles, using embryonic brain preparations from turtle, various lizards and other squamates, as well as crocodiles and *Sphenodon*. Based on these observations, embryonic brain development was studied in more details in the turtle (*Emys orbicularis*) and lacertilian lizards (*Lacerta agilis* and *Lacerta trilineata*) using electron microscopy, Golgi impregnations and [3H]-thymidine autoradiography. These two species were selected based on histological observations which, in agreement with anatomical and palaeontological data, suggested that among the reptiles, turtles and lizards have, respectively, a rudimentary and a highly evolved type of cortical architectonic organization.

The radial organization of the cortical plate as an homoplastic character

We attempted to correlate the observed developmental differences in the cortical plates of *Emys* and *Lacerta* with the lineage of reptiles and their relation to mammalian ancestors. Unfortunately, many aspects in of reptilian evolution remain obscure. The monophyletic origin of amniotes is generally accepted. A common reptilian ancestor (probably during the Pennsylvanian) gave rise to several independent lines, some of which led to living reptiles, mammals and birds. The textbook view is that a first branch, the synapsids, separated early and gave rise to mammals; a second branch led to chelonians, via a poorly understood lineage; a second branch led to Rynchocephalia (of which *Sphenodon punctatus* is the only living representative) and to lizards and their ophidian derivatives; and a third branch gave rise to the crocodilians (via thecodonts) and to birds (via saurischian dinosaurs). The anapsid lineage, represented by living turtles (chelonians) was until recently considered an individual phylum, but this is now controversial. Recent palaeontological and molecular data (Rieppel & deBraga 1996, Hedges & Poling 1999) suggest that the long-held assumption that turtles are a model for primitive amniote organization may not be correct, and the position of *Sphenodon* is also questionable. From a neuroembryological standpoint, however, the turtle cortex can certainly be described as rudimentary in comparison to other reptiles and mammals (Goffinet 1983, 1992, Goffinet et al 1986).

Comparative observations of cortical plate development suggest that the radial organization of the cortical plate is not an all-or-nothing phenomenon and that, among reptiles, *Emys* and *Lacerta* represent two extremes of this cytological feature (Fig. 1). While the turtle cortical plate appears rudimentary in terms of laminar definition and radial neuronal organization, the layering of most cortical areas in lacertilians is sharply defined and its constituent neurons assume an elaborate radial organization. The cortical plate of crocodilians is prominent; its neuronal density is not as dense as in lacertilians, but it is still better organized than in turtles. In *Sphenodon*, the architectonics of the cortical plate are elaborate at the level of the hippocampus, whilst the dorsal pallium appears poorly organized. In snake embryos, the cytoarchitectonics in the various sectors of the cortical plate appear more highly organized than in turtles but less than in lizards.

The radial orientation of developing cortical neurons is thus expressed differently in the various reptilian phyla (Goffinet 1983). A first possibility is that radial cortical organization was present in stem reptiles, but has been lost progressively and to various extents in the different lineages. Given the importance of radial organization for cortical development in mammals and its alteration as a result of reelin or Dab1 mutations, we regard this view as unlikely. We consider it more reasonable to suggest that radial organization has been acquired independently, gradually and to variable extents after phyletic divergence. It would then provide an example of homoplasy, due to evolutionary convergence, for it 'involves the independent evolution of similar characters in organisms possessing distant common ancestry' (Northcutt 1981). As in other cases of homoplasy, it reveals that 'similar solutions to biological problems have occurred independently' (Northcutt 1981) and probably correspond to efficient solutions.

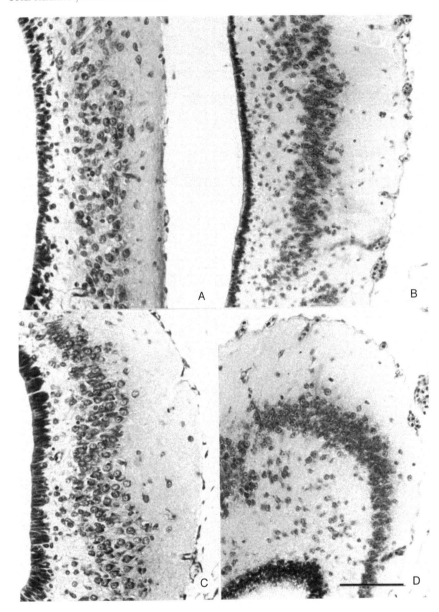

FIG. 1. Comparative architectonics of the dorsal (A, B) and medial (C, D) cortex in *Emys orbicularis* (A, C) and *Lacerta agilis* (B, D) at a mid-incubation stage. The species difference in radial organization of the cortical plate is particularly evident. Other species, for example crocodilians and *Sphenodon*, have a cortex with a radial organization that can be described as intermediate between turtles and lizards. Adapted from Goffinet (1983). Bar = 100 μm.

Corticogenesis in reptiles proceeds from outside to inside

In order to study corticogenesis in reptiles, and specifically to assess whether there is any correlation between radial architectonics and the inside-out gradient in mammals, tritiated thymidine autoradiographic studies was carried out in turtle and lacertilians. In the lateral and dorsal cortices of both species, radial histogenesis follows an outside-to-inside gradient (Fig. 2), i.e. late-generated neurons settle at deeper levels in the cortex than older cells. The situation is slightly different at the level of the medial cortex, where subtle differences

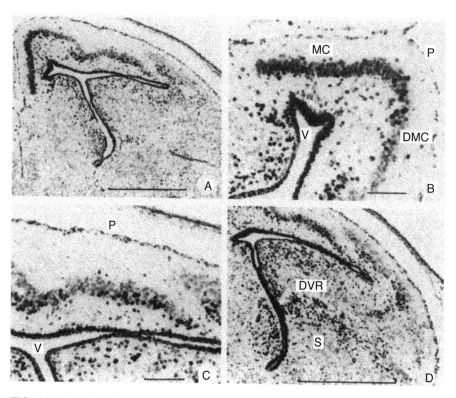

FIG. 2.   Examples of tritiated thymidine autoradiographic studies of neurogenesis in the lacertilian brain (adapted from Goffinet et al 1986). (A) Overview of a frontal section with labelled, comparatively younger neurons and unlabelled, older neurons. The lateral to medial gradient of cortical neurogenesis is evident. Bar = 500 μm. (B) In the medial cortex, no neurogenetic gradient is seen in the radial dimension. Bar = 100 μm. (C) In the dorsal cortex, early-generated neurons (unlabelled) are clearly located externally, with younger neurons (labelled) at inner level, i.e. the gradient is directed from outside to inside. Bar = 100 μm. (D) In the lateral cortex, the same outside-inside gradient as in the dorsal cortex is found. The dorsal ventricular ridge (DVR) and striatum (S) appear as two separate neurogenetic compartments. Bar = 500 μm. DMC, dorsomedial cortex; MC, medial cortex; P, pial surface; V, ventricle.

between the two species are found. In turtles, an outside-to-inside gradient is seen in the medial, as in the lateral and dorsal cortices, whereas it is impossible to define any radial gradient in the lacertilian medial cortex. Rather surprisingly, since our study was completed, no other analysis of the timing of corticogenesis has been performed in any other reptilian species, so that our knowledge remains fragmentary (Goffinet 1992, Goffinet et al 1986). Our data suggest, however, that reptilian corticogenesis proceeds from outside to inside in most if not all reptilian cortices, independently from the degree of radial architectonics. The acquisition of an 'inverted', inside-to-outside pattern of corticogenesis can thus be considered a specific step in the evolution of the cerebral cortex from the stem reptile to the mammalian stage (synapsid radiation).

As already mentioned above, when radial cortical organization is disturbed by mutations in reelin or Dab1, corticogenesis no longer follows the normal, 'inverted' pattern, but instead proceeds from outside to inside, as in reptiles (Caviness 1982, Gonzales et al 1997). Altogether, these observations suggest that radial neuronal organization is necessary but not sufficient for cortical development to proceed from inside to outside. Radial architectonics and inside-outside maturation would then define two sequential steps of cortical evolution in the synapsid lineage (Lambert de Rouvroit & Goffinet 1998a).

reelin/Dab1 expression

The cloning of reelin (D'Arcangelo et al 1995) and Dab1 (Howell et al 1997, Sheldon et al 1997, Ware et al 1997) provided an opportunity to study comparatively their expression during cortical development in mammals and reptiles using *in situ* hybridization. The necessary species-specific probes were generated by cloning partial sequences from reelin and Dab1 in a turtle (*E. orbicularis*), a lizard (*Lacerta viridis*) and a bird (chick) in order to study their expression by *in situ* hybridization (Bernier et al 1998). We have thus far been unable to obtain sequences from crocodilians and *Sphenodon*. In addition, monoclonal antibody (mAb) 142, raised against mouse reelin (de Bergeyck et al 1998), reacts well with reelin from humans, lizard and several other species, allowing comparative studies of reelin protein expression.

CLUSTAL alignments of the 340 C-terminal amino acids of reelin (3461 amino acids in length) in humans, mice, chicks, lizards and turtles show that the turtle reelin clusters with that of birds. Although preliminary, this observation fits with recent observations that turtle protein sequences are more similar to archosaurs than to other groups (Hedges & Poling 1999). Work is in progress to clone larger segments of reelin and Dab1 in these and other species. Although, in all species studied, reelin and Dab1 expression is clearly not restricted to cerebral cortical areas, only cortical expression will be considered here.

In all mammals studied thus far, at least prior to birth, reelin-positive cells are restricted to the marginal zone (D'Arcangelo et al 1995, Ogawa et al 1995, Schiffmann et al 1997). They are abundant and express high concentrations of reelin. These cells include Cajal-Retzius neurons but also some other, less well-characterized elements, and the discussion of this point is beyond this chapter (see Meyer et al 1998, Meyer & Goffinet 1998). In turtles, the reelin-related signal is also found mainly in the marginal zone of all cortical sectors, but also, although to a lesser extent, in the cortical plate itself (Fig. 3). In the turtle marginal zone, reelin expression is associated with large, strongly positive cells that are quite rare, scattered over all cortical areas, but more abundant in medial than in dorsal and lateral regions. These disperse reelin-positive cells might be homologous to mammalian Cajal-Retzius cells. In chick cortex, reelin-positive cells also are similarly restricted to the marginal zone, but are even rarer than in turtle cortex. Preliminary results using mAb142 in crocodile embryos suggest that the pattern of reelin expression is comparable, restricted to large, scattered horizontal neurons intrinsic to the marginal zone (G. Meyer & A. M. Goffinet, unpublished results 1999). In lizard cortex, by contrast, reelin expression is different. In the lateral cortex, reelin expression is found all over. In other cortical areas (medial, dorsal and dorsomedial), reelin expression is found in two layers bracketing the cortical plate. First, it is associated with large, disperse horizontal neurons in the marginal zone that are quite reminiscent to those in turtle. Second, there is strong reelin expression in more abundant neurons below the lizard cortical plate (Figs 4 and 5).

In the mammalian cortex, the pattern of Dab1 expression was shown to be complementary to that of reelin, in that Dab1 was expressed in all cortical plate neurons, whereas reelin is found solely in Cajal-Retzius cells (Rice et al 1998). With the exception of the lizard lateral cortex, in which the reelin and Dab1 expression patterns overlap, the pattern described in rodents is also found in reptiles, with Dab1 being clearly expressed in the cortical plate in chick, turtle and lizard (Fig. 4). Although the data are still incomplete, they suggest that the complementarity of reelin and Dab1 expression is a rather general feature. These observations are in line with the hypothesis that reelin produced by Cajal-Retzius and other cells is secreted in the extracellular matrix and acts at a distance on target neurons, instructing them to assume their architectonic differentiation, and the response of target cells requires a normal Dab1 gene product (Goffinet 1997, Lambert de Rouvroit & Goffinet 1998a,b). During cortical evolution, important changes occurred in the number of reelin-positive cells and in their individual level of reelin expression.

A model of cortical evolution

The data summarized above suggest the following model. A cerebral cortex was probably present in stem reptiles and had rudimentary architectonic features that

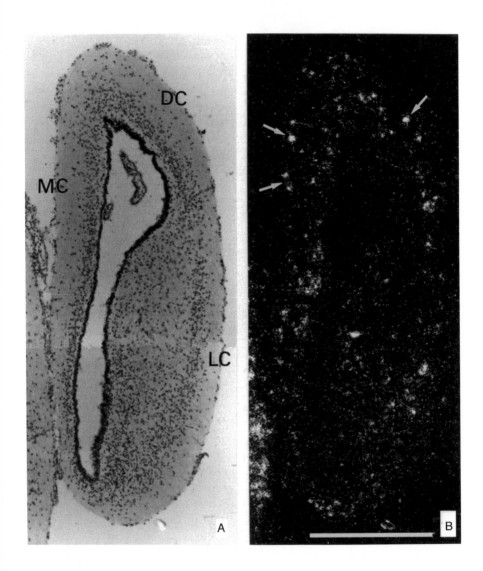

FIG. 3.   Reelin mRNA expression in turtle embryonic cortex. Coronal section through a turtle cortex at Yntema's stage 23 (Yntema 1968), in bright (A) and darkfield (B) view. Note the poor radial organization of the cortical plate, in which there appears to be low expression of reelin mRNA (stars on the darkfield panel B). By contrast, there are strongly positive cells (arrows) in the marginal zone of the dorsal cortex (DC) and medial cortex (MC). No reelin-positive cells are seen below the cortical plate. Bar = 500 µm. LC, lateral cortex.

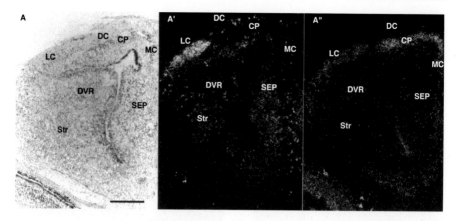

FIG. 4. Reelin and Dab1 mRNA expression in *Lacerta* at Dufaure and Hubert's stage 35 (Dufaure & Hubert 1961). (A) Hematoxylin-eosin-stained section, showing anterior forebrain with lateral (LC), dorsal (DC) and medial (MC) cortex, as well as anterior dorsal ventricular ridge (DVR), striatum (Str), and septal nuclei (SEP). (A') *In situ* hybridization for reelin mRNA, which is expressed rather diffusely in the lateral cortex, while reelin expression is found in two layers, in large cells in the marginal zone and subplate, and at the level of the dorsal and medial cortex. (A'') *In situ* hybridization for Dab1 mRNA, shown at a comparable level. Dab1 expression is diffuse in lateral cortex, but restricted to the cortical plate (CP) in dorsal and medial cortical areas. At this level, the zone of Dab1 expression corresponds to the negative area of reelin expression (panel A'). Bar = 200 μm.

are perhaps best approximated by the turtle cortex. In this primitive cortex, the three main divisions were already present, radial neuronal organization was poorly developed and the histogenetic gradient was directed from outside to inside. During subsequent evolution, the cortical plate increased both in terms of cell number and architectonic organization. Our data allow the definition of at least two successive acquisitions. First, radial organization developed gradually, leading to a cortical organization that would be illustrated today by the lacertilian cortex. Histogenesis in this cortical plate still proceeded from outside to inside. Second, the inverted gradient, from inside to outside, was acquired, leading to the present organization of the mammalian cortical plate.

Support for this model is provided by recent observations of cyclin-dependent kinase 5 (*Cdk5*) (Gilmore et al 1998) and *p35* knockout mice (Chae et al 1997, Kwon & Tsai 1998). In these mutant animals, the inside-outside gradient of cortical histogenesis is disturbed while radial cortical organization is relatively preserved. In this restricted sense, the cortex of the *Cdk5* and *p35* knockout mice is reminiscent of that in lizards, suggesting that: (1) reelin and Dab1 are necessary for the development of radial organization and of a cell-poor marginal zone; and (2) they are necessary but not sufficient for the development of the inside-out gradient that requires other, mostly unidentified factors, for example a sufficient number of

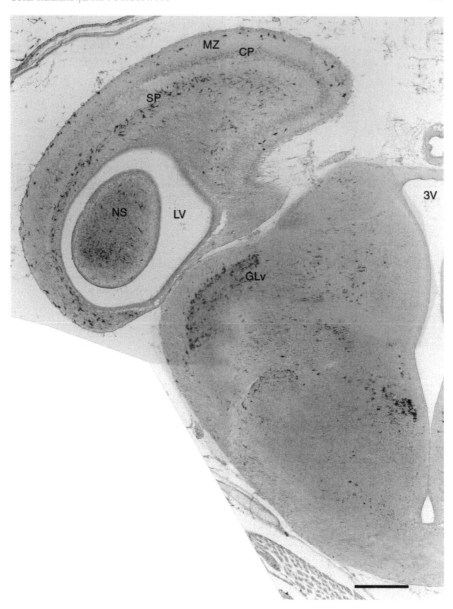

FIG. 5. Reelin protein expression in lizard embryonic forebrain. The reelin protein is revealed with mAb142. In the hemisphere, there is clearly no reelin protein in the cortical plate (CP) but reelin is found in large cells in the marginal zone (MZ) and below the cortical plate in the subplate (SP). Reelin-positive cells are also found in the nucleus sphaericus (NS) and in various diencephalic nuclei that encompass particularly the reticular thalamic components such as the ventral lateral geniculate nucleus (GLv). 3V, third ventricle; LV, lateral ventricle. Bar = 200 μm.

neurons. If this is correct, reelin and Dab1 on the one hand, and Cdk5 and p35 on the other hand, must act in sequence to generate the inside-out gradient (Lambert de Rouvroit & Goffinet 1998b). Most probably, several other proteins are also involved in this complex morphogenetic process.

This model obviously provides barely a sketchy outline of an exceedingly complex phenomenon and needs to be put to the test. In this regard, it is necessary to examine reelin and Dab1 expression during brain development in crocodilians, *Sphenodon* and at least a few other species. This should be completed by studies of Cdk5, p35 and of the various factors implicated in cortical development and differentiation, the list of which increases by the year. Clearly, the 'evo-devo' approach of the cerebral cortex is still in its infancy and remains largely open to future investigation.

Acknowledgements

I wish to thank C. Lambert de Rouvroit, G. Meyer, C. Pieau and A. Raynaud for help and discussion. Work described in this paper was carried out with support from the Fonds National de la Recherche Scientifique, Fonds de la Recherche Scientifique Médicale, Actions de Recherches Concertées and Fondation Médicale Reine Elisabeth, all from Belgium.

References

Bernier B, Goffinet AM, Lambert de Rouvroit C 1998 Comparative studies of reelin during brain development. Eur J Neurosci S10:381

Caviness VS 1982 Neocortical histogenesis in normal and *reeler* mice: a developmental study based upon ^{3}H-thymidine autoradiography. Dev Brain Res 4:297–326

Caviness VS, Rakic P 1978 Mechanisms of cortical development: a view from mutations in mice. Annu Rev Neurosci 1:297–326

Chae T, Kwon YT, Bronson R, Dikkes P, Li E, Tsai LH 1997 Mice lacking p35, a neuronal specific activator of Cdk5, display cortical lamination defects, seizures, and adult lethality. Neuron 18:29–42

D'Arcangelo G, Miao GG, Chen SC, Soares HD, Morgan JI, Curran T 1995 A protein related to extracellular matrix proteins deleted in the mouse mutant *reeler*. Nature 374:719–723

de Bergeyck V, Naerhuyzen B, Goffinet AM, Lambert de Rouvroit C 1998 A panel of monoclonal antibodies against Reelin, the extracellular matrix protein defective in *reeler* mutant mice. J Neurosci Meth 82:17–24

Dufaure JP, Hubert J 1961 Table de développement du lézard vivipare, *Lacerta vivipare*. Arch Anat Microsc Morphol Exp 50:309–327

Gilmore EC, Ohshima T, Goffinet AM, Kulkarni AB, Herrup K 1998 Cyclin-dependent kinase 5-deficient mice demonstrate novel developmental arrest in cerebral cortex. J Neurosci 18:6370–6377

Goffinet AM 1983 The embryonic development of the cortical plate in reptiles: a comparative study in *Emys orbicularis* and *Lacerta agilis*. J Comp Neurol 215:437–452

Goffinet AM 1992 The *reeler* gene: a clue to brain development and evolution. Int J Dev Biol 36:101–107

Goffinet AM 1997 Developmental neurobiology. Unscrambling a disabled brain. Nature 389:668–669

Goffinet AM, Daumerie C, Langerwerf B, Pieau C 1986 Neurogenesis in reptilian cortical structures: ^{3}H-thymidine autoradiographic analysis. J Comp Neurol 243:106–116

Gonzales JL, Russo CJ, Goldowitz D, Sweet HO, Davisson MT, Walsh CA 1997 Birthdata and cell marker analysis of *scrambler*: a novel mutation affecting cortical development with a *reeler*-like phenotype. J Neurosci 17:9204–9211

Hedges SB, Poling LL 1999 A molecular phylogeny of reptiles. Science 283:998–1001

Howell BW, Hawkes R, Soriano P, Cooper JA 1997 Neuronal position in the developing brain is regulated by mouse *disabled-1*. Nature 389:733–737

Kwon YT, Tsai LH 1998 A novel disruption of cortical development in *p35*(-/-) mice distinct from *reeler*. J Comp Neurol 395:510–522

Lambert de Rouvroit C, Goffinet AM 1998a The *reeler* mouse as a model of brain development. Adv Anat Embryol Cell Biol 150:1–106

Lambert de Rouvroit C, Goffinet AM 1998b A new view of early cortical development. Biochem Pharmacol 56:1403–1409

Meyer G, Goffinet AM 1998 Prenatal development of reelin-immunoreactive neurons in the human neocortex. J Comp Neurol 397:29–40

Meyer G, Soria JM, Martínez-Galán JR, Martín-Clemente B, Fairén A 1998 Different origins and developmental histories of transient neurons in the marginal zone of the fetal and neonatal rat cortex. J Comp Neurol 397:493–518

Northcutt RG 1981 Evolution of the telencephalon in non-mammals. Annu Rev Neurosci 4:301–350

Ogawa M, Miyata T, Nakajima K et al 1995 The *reeler* gene-associated antigen on Cajal-Retzius neurons is a crucial molecule for laminar organization of cortical neurons. Neuron 14:899–912

Rice DS, Sheldon M, D'Arcangelo G, Nakajima K, Goldowitz D, Curran T 1998 Disabled-1 acts downstream of Reelin in a signaling pathway that controls laminar organization in the mammalian brain. Development 125:3719–3729

Rieppel O, deBraga M 1996 Turtles as diapsid reptiles. Nature 384:453–455

Schiffmann SN, Bernier B, Goffinet AM 1997 Reelin mRNA expression during mouse brain development. Eur J Neurosci 9:1055–1071

Sheldon M, Rice DS, D'Arcangelo G et al 1997 *Scrambler* and *yotari* disrupt the *disabled* gene and produce a *reeler*-like phenotype in mice. Nature 389:730–733

Sheppard AM, Pearlman AL 1997 Abnormal reorganization of preplate neurons and their associated extracellular matrix: an early manifestation of altered neocortical development in the *reeler* mutant mouse. J Comp Neurol 378:173–179

Ware ML, Fox JW, González JL et al 1997 Aberrant splicing of a mouse *disabled* homolog, *mDab1*, in the *scrambler* mouse. Neuron 19:239–249

Yntema CL 1968 A series of stages in the embryonic development of *Chelydra serpentina*. J Morphol 125:219–251

DISCUSSION

Reiner: I guess you are in a position to say that the reelin/Dab1 system is necessary for generating the inside-out gradient but not sufficient, because turtles have Cajal-Retzius cells that express reelin in their cortex but they still have the outside-in gradient.

Goffinet: Yes, that's the idea.

Reiner: What more would a turtle need? Do they need more migratory cells?

Goffinet: It's anyone's guess. They probably do need more migratory cells.

Rakic: It may not be too surprising that one molecule, i.e. reelin, is used in different ways in different structures or developmental contexts in evolution. This is also the case for classical neurotransmitters such as glutamate, which may act as a trophic factor or morphogen. For example, glutamate controls the rate of neuronal migration before the establishment of synaptic connections (Komuro & Rakic 1993).

Pettigrew: Are there other structures apart from the cerebral and cerebellar cortex in which reelin plays a role?

Goffinet: There are many different kinds of laminations, some of which are reelin-dependent and others reelin-independent. Reelin-dependent lamination occurs in the hippocampus, cortex and cerebellum, i.e. in all the structures that are defective in reelin-negative mice. Examples of laminated structures in which reelin is of little or no importance include the olfactory bulb and the retina.

Rubenstein: What about the superior colliculus and the spinal cord?

Goffinet: The superior colliculus has reelin-dependent lamination. The spinal cord is slightly abnormal in reelin-negative mice, but the defects are subtle.

Levitt: Joe Yip of the University of Pittsburgh, in preliminary studies, has also shown that in reelin-negative mice there are also migratory abnormalities in the region of the pre-ganglionic neurons in the spinal cord (personal communication 1999).

Goffinet: I'm sure that if you look carefully at every region in the brain in reelin-negative mice, you will find some kind of abnormalities. There are probably even subtle abnormalities in the retina and the olfactory bulb, although in these structures lamination is largely reelin-independent.

Pettigrew: I would also like to make a comment about the use of DNA data. There are quite a few cases now where preposterous phylogenies have been generated from large maps of DNA data. It's not a question of there being lack of data, rather that the DNA data can be quantitatively misleading. The two classic examples of this are in *Dictyostelium*, which is a eukaryote but the computer programs say, with $P = 10^{-10}$, is a prokaryote, and secondly *Amphioxus*, which is a cephalochordate but the computer programs say it is outside the echinoderms. In both of those cases we can show that the error comes about because of base compositional bias. Naylor & Brown (1998) have also shown that, in some cases, even amino acids can give the wrong answer.

Goffinet: The DNA data should only be used on the same footing as phenotypic markers, and there are no hierarchies.

Evans: There is some evidence that DNA data work better when you are looking at taxa that are closely related to one another.

Pettigrew: Is there a reason why DNA can't converge just as beautifully as neural structures and morphological structures can?

Goffinet: I thought about this when I was comparing the sequences. I found that the sequences of the coding regions were all similar, so perhaps we will obtain more useful information by comparing the promoter sequences.

Papalopulu: That's a good point. The question is, why are these brains so different if they have so many similar genes? I would argue that in order to make a large difference in morphology, you only need subtle differences in gene sequence, but those differences are likely to change the timing of expression. For example, if a gene involved in the regulation of the cell cycle or cell death is switched on for a little longer, it is likely to have dramatic effects on brain morphology. For example, if a gene product of a particular gene feeds back and switches off its own promoter, it will switch itself off after it reaches a certain level of expression. However, if the promoter is made a little less sensitive, then the gene will be switched on for a little longer. This is a small change, but it can have a profound effect. When we look at the expression of genes in different organisms, we tend not to compare the expression levels and how long they remain switched on, so it's possible that the same genes may be expressed in a similar pattern but give rise to different outcomes.

Levitt: I will also show in my presentation that this is also the case for signal transduction molecules. If you raise the receptor expression levels slightly, you can completely change the fate of the cell.

Herrup: I have two questions. First, what is the nature of the diffuse staining in your *in situ* hybridizations, because it seems qualitatively different from the more focused staining in what you're calling the Cajal-Retzius cell? And second, if reelin is expressed essentially all over the nervous system, why don't pyramidal cells end up in the thalamus?

Goffinet: Part of the diffuse signal is just the background signal, although I agree there is some diffuse staining in the cortical plate, and we don't yet know if this low-level expression is important. Your second question is more difficult to answer. I'm sure we will find that this pattern of diffferent reelin expression is present in all vertebrates.

Molnár: Is reelin present in *Drosophila*?

Goffinet: No, it's not present in invertebrates, as far as we can tell.

Molnár: How do the expression patterns of reelin and Dab1 differ in the dorsal ventricular ridge (DVR)?

Goffinet: This is difficult to analyse. We are trying to do double *in situ* hybridizations to see if they co-localize in the lateral cortex. We know that they do not co-localize in the cortex, whereas in the DVR there is some overlap.

Broccoli: How far can reelin diffuse in the extracellular matrix? And is it possible that differences in this distance between species could be important in an evolutionary sense?

Goffinet: It's an important point, but I don't know the answer. It's complicated by the fact that reelin is processed by a metalloprotease, probably after secretion.

Rubenstein: In the mammalian nervous system, are there other structures besides the cerebral cortex that are inside-out, and are there other structures in any species that are inside-out, because we may be able to use these as models?

Goffinet: I don't know. There is the huge vagal lobe, but this hasn't been studied in terms of neurogenetic gradients.

Karten: In the chick tectum, lamination doesn't follow a simple inside-out pattern. But if you look at the data in a different way, the inside-out pattern is not the primary determinant. Thus, in the spinal cord motor neurons form first, dorsal neurons form second and interneurons form third. If you look at the cortical plate, the cells that form the deepest layers are the oldest, they go from VI, V, IV, III and II so that you are actually looking at the efferents first, the recipients next, and the interneurons last. Therefore, the inside-out pattern may just be an epiphenomenal property of cortex, and the true determinants of sequence of neurogenesis may be where they sit in relation to other properties, such as their connections.

References

Komuro H, Rakic P 1993 Modulation of neuronal migration by NMDA receptors. Science 260:95–97

Naylor G JP, Brown WM 1998 Amphioxus mitochondrial DNA, chordate phylogeny, and the limits of inference based on comparison of sequences. Syst Biol 47:61–77

The contribution of the ganglionic eminence to the neuronal cell types of the cerebral cortex

J. G. Parnavelas, S. A. Anderson*, A. A. Lavdas, M. Grigoriou, V. Pachnis and J. L. R. Rubenstein*

*Department of Anatomy and Developmental Biology, University College London, London WC1E 6BT, UK, and *Nina Ireland Laboratory of Developmental Neurobiology, Department of Psychiatry, University of California, San Francisco, CA 94143, USA*

Abstract. The principal neuronal types of the mammalian cerebral cortex are the excitatory pyramidal cells and the inhibitory interneurons, the non-pyramidal cells. It is thought that these neurons arise in the ventricular zone surrounding the telencephalic ventricles. From there, newly generated neurons migrate outward along the processes of radial glial cells to reach the cortical plate where they accumulate in an 'inside-out' sequence to form the six-layered structure of the neocortex. Here we review emerging evidence that pyramidal neurons are generated in the cortical ventricular zone, whereas the majority of the non-pyramidal cells arise in the ganglionic eminences of the ventral telencephalon. These neurons follow tangential migratory routes to reach their positions in the developing cortex.

2000 Evolutionary developmental biology of the cerebral cortex. Wiley, Chichester (Novartis Foundation Symposium 228) p 129–147

The mammalian neocortex is divided into areas that were originally defined in terms of their cytological differences and later discovered to serve different functions. All areas share a common basic structure, with neurons arranged in six layers oriented tangentially (i.e. parallel to the surface of the cortex). The majority of cortical neurons (70–80%; Rockel et al 1980, Parnavelas et al 1989) are pyramidal cells present in all layers except layer I. These are the projection cells of the cortex that utilize the excitatory amino acid L-glutamate as a neurotransmitter. The rest of the cortical neurons, scattered in all layers, are the non-pyramidal cells. These are the cortical interneurons that contain the inhibitory neurotransmitter γ-aminobutyric acid (GABA; Parnavelas et al 1989). An important issue in understanding the development of the cortex is where, when and how the diversity of its neuronal cell types is specified during ontogenesis.

It is widely thought that neurons of the cerebral cortex arise in the germinal ventricular zone. Postmitotic neurons migrate outward to the cortical plate where they accumulate in an 'inside-out' sequence to form the characteristic six-layered structure of the neocortex (Rakic 1988). The earliest-born cells are an exception to this rule as they accumulate at the outer edge of the cerebral wall, before the cortical plate appears, to form the preplate. This layer is subsequently split by the arriving cortical plate neurons into the superficial marginal zone (cortical layer I) and the subplate below the cortical plate (Uylings et al 1990). The origin of the cells of the preplate, the Cajal-Retzius cells of the marginal zone and of the subplate neurons, has not been well documented, although it is thought that they originate in the telencephalic ventricular zone (Marín-Padilla 1998). Evidence suggests that these early neurons have important functions during development: Cajal-Retzius cells appear to play a role in neuronal migration and layer formation (Ogawa et al 1995, Frotscher 1998), and subplate neurons are involved in the establishment of cortical connections (McConnell et al 1994).

The cortical plate, sandwiched between the two components of the preplate, steadily thickens as neurons migrate through the subplate and take up positions under the marginal zone. The migration of neurons to the cortical plate is guided by the processes of radial glia that span the entire thickness of the developing cortex (Rakic 1988). However, both *in vivo* and *in vitro* experiments have suggested that migrating neurons may ignore radial glial fibres and adopt tangential migratory paths to their positions in the cortical plate. Support for tangential migration has come from tracing studies of DiI (1,1'dioctadecyl-3,3,3',3'-tetramethylindocarbocyanine perchlorate)-labelled postmitotic neurons (O'Rourke et al 1995), and from lineage analyses with retroviruses. These lineage studies have shown that clonal relatives may be located at great distances along the rostrocaudal axis of the cortex, although they were supposedly generated at a focal point in the ventricular zone (Walsh & Cepko 1992). Radial and non-radial migratory routes may expose young neurons to different cues that may be important for the acquisition of specific phenotypes. Experiments in the chick optic tectum have provided evidence that different migratory paths are related to phenotypic choices of clonally related cells (Gray & Sanes 1991). Cell type-specific migratory routes may be a selective mechanism to sort out different phenotypes generated by multipotential progenitors.

Earlier lineage studies (Parnavelas et al 1991, Luskin et al 1993, Mione et al 1994) suggested that the two neuronal populations in the cortex, the pyramidal and non-pyramidal cells, arise from different progenitors in the ventricular zone. It is now known that not all clonally related cells maintain the spatial relationship that they had before migration (Walsh & Cepko 1993), but it is noteworthy that those that do invariably show the same phenotype (Reid et al 1995). We have recently used a lineage marker in combination with BrdU to analyse the pattern of generation of pyramidal and non-pyramidal cell types (Mione et al 1997). We found that only pyramidal neurons maintain a close spatial relationship with their clonal relatives in the cortex, which can be achieved through radial migration. In contrast, labelled non-pyramidal neurons were found as isolated cells or as pairs of clonally related neurons. Their low content of BrdU indicated that these cells were part of larger clones, and suggested that their isolation was the result of non-radial (tangential) migration through the cortex. These findings pose two alternative interpretations: either clonally related cells, instructed to develop a particular phenotype, are also endowed with the ability to use a specific migratory pathway; or cues encountered during radial and tangential migration are responsible for the pyramidal and non-pyramidal phenotypes. Recent work by Tan et al (1998) has provided evidence for the former possibility with regards to the generation and migration of pyramidal neurons. Using highly unbalanced mouse embryonic stem cell chimeras, they confirmed the results of our lineage analysis; they found that radially dispersed neurons contained glutamate, the neurochemical signature of pyramidal neurons, whereas tangentially dispersed cells were predominantly GABAergic. Their study demonstrated, consistent with earlier results from transgenic mouse lines (Soriano et al 1995), that specification of the pyramidal lineage does occur at the level of the progenitor, before the onset of neurogenesis. The association of radial migration with pyramidal neurons would indicate that this neuronal type, which constitutes the majority of neocortical neurons, may provide the vehicle for relaying the proto-map as proposed by Rakic (1988).

Our understanding of the origin and development of non-pyramidal cell lineages is not at all clear. Our lineage studies (Mione et al 1997) suggested that non-pyramidal cells, scattered in the cortex as isolated neurons or pairs of clonally related cells, were part of larger clones. This raised the question as to whether there exist in the cortical ventricular zone progenitors committed to producing only non-pyramidal cells. The analysis of chimeric mice by Tan et al (1998) raised the same question, and concluded that whether these neurons are generated from progenitors with single or mixed potential, they have the tendency to be diffusely scattered in the cortex. Even though non-pyramidal cells are not as numerous in the neocortex as pyramidal neurons (20–30% of the neuronal population), the same cortical space contains much larger pyramidal clones, numbering up to 30 cells after intraventricular injections of retrovirus in embryos at embryonic day (E) 14 (Parnavelas et al 1991, Mione et al 1994). These observations point to another source of cortical interneurons.

Ganglionic eminence: a source of cortical interneurons

Lateral ganglionic eminence

Sources of neurons destined for the neocortex have been discovered in the ganglionic eminences of the ventral telencephalon. Porteus et al (1994) reported that cells expressing *Dlx2* appear to migrate out of the ventral telencephalon into the developing cerebral cortex. More recently, De Carlos et al (1996) and Tamamaki et al (1997) provided unequivocal evidence that cells in the lateral ganglionic eminence (LGE), the primordium of the striatum, transgress the corticostriatal boundary as they migrate into the developing neocortex, and are distributed predominantly in the intermediate zone. We have utilized the fluorescent tracer DiI to investigate the migration and disposition in the cortex of neurons originating in the LGE, and have characterized the identity of these neurons using cell-specific markers (Anderson et al 1997a). Using slice cultures of embryonic mouse brains, we found that placement of DiI in the LGE as early as E11.5 resulted in the presence of labelled cells in the intermediate zone and cortical plate of the developing cortex 36 h later. Migrating cells, bearing a thick leading process in the direction of the migration, appeared to round the corticostriatal sulcus and were directed tangentially toward the neocortex. Double-labelling experiments with antibodies against GABA or calcium-binding proteins clearly showed that a substantial proportion of the DiI-labelled cells expressed GABA or calbindin. Furthermore, the number of GABA-expressing cells in neocortical slices was reduced significantly when the neocortex was separated from the ventral telencephalon.

In an attempt to characterize the molecular mechanisms that regulate the migration of neurons from the LGE to the cortical plate and intermediate zone, we focused on the transcription factors *Dlx1* and *Dlx2*, which are homeobox-containing genes with virtually identical patterns of expression in the developing forebrain (Bulfone et al 1993, Porteus et al 1994). These genes are expressed in the developing cortex from E13.5 and, like GABA and calbindin, the expression of *Dlx1* is reduced in the neocortex of transected slice cultures. Double-labelling for *Dlx1* and either GABA or calbindin revealed coexpression in tangentially oriented cells in the intermediate zone resembling the neurons that originate in the LGE (Anderson et al 1997a). This observation suggests that the *Dlx* genes may be required for cell migration from the ventral telencephalon. Analysis of mice with a mutation in both *Dlx1* and *Dlx2* has provided evidence in support of this suggestion. Thus homozygous *Dlx1/2* mutants showed abnormal migration and accumulation of partially differentiated cells in the LGE (Anderson et al 1997b; Fig. 1). Furthermore, these mutants showed a significantly reduced number of GABA- and calbindin-expressing cells in the neocortex (Fig. 2). Analysis of the cortical lamination of the *Dlx1/2* mutants,

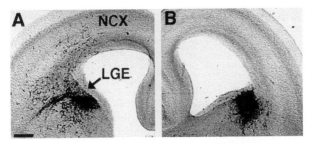

FIG. 1.   Comparison of cell migration out of the lateral ganglionic eminence (LGE) in slice cultures from wild-type (A) and $Dlx1/2^{-/-}$ mouse mutant (B). Slices, from embryonic day (E) 15.5 mice, were cultured for 36 h. Cells in the wild-type slice (A) have migrated from the LGE into the neocortex (NCX). This migration was absent in the mutant slice (B). The DiI (1,1'dioctadecyl-3,3,3',3'-tetramethylindocarbocyanine perchlorate) was photoconverted into diaminobenzidine. Bar $= 200\,\mu$m (from Anderson et al 1997a).

using BrdU labelling, indicated that the radial migration of the cortical pyramidal neurons is unaffected.

Medial ganglionic eminence

A similar experimental approach was applied to investigate the contribution of the medial ganglionic eminence (MGE) to the cellular composition of the developing neocortex (Lavdas et al 1999). Thus, DiI crystals were placed in the MGE of cultured slices prepared from rat embryos between stages E13 and E19. After two days *in vitro*, slices prepared from E13 and E14 embryos displayed labelled neurons emerging from the MGE. A number of these cells were directed ventrolaterally, others were found rounding the corticostriatal sulcus and directed either dorsolaterally or toward the temporal cortex, while others appeared to reach the most superficial aspect of the developing cortex, the preplate, and were oriented parallel to the pial surface (Fig. 3A). The tangentially oriented cells showed features typical of Cajal-Retzius cells according to earlier descriptions in the cortex of rat embryos (Bradford et al 1977, Derer & Derer 1990). Slices prepared from E15 and E16 embryos contained migrating cells not only in the marginal zone, but also in the lower intermediate zone and in the subplate; a small number of cells was also noted in the cortical plate (Fig. 3B,C). These cells typically had a long and thick leading process in the direction of the migration (Fig. 3D). A similar group of cells in the intermediate zone and subplate were seen in slices injected with DiI at E17, but these preparations did not show any labelling in the marginal zone. Slices prepared from older embryos, E18 and E19, did not show any labelling in the neocortex. Migration of labelled MGE cells within the cortex was, for the most part, along tangentially oriented

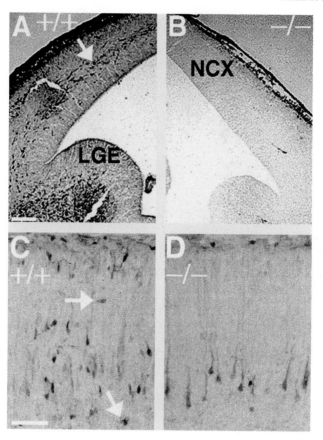

FIG. 2. Analysis of the developing neocortex (NCX) in the *Dlx1/2* mutants. Immunohistochemistry for calbindin (A to D) in coronal sections. (A) At embryonic day (E) 14.5 numerous tangentially oriented, calbindin-reactive cells are present in the intermediate zone (arrow) of the wild-type neocortex. (B) The number of these cells is markedly reduced in the mutant. (C and D) At birth, calbindin-expressing cells are present in layer V (pyramidal neurons) and in layer I of the mutant section (D), but far fewer calbindin-positive interneurons (arrows in C) are present in the mutant cortical plate. Bars = 100 μm (from Anderson et al 1997a).

routes; and, like cortical neurons arising in the LGE, they expressed GABA. Further characterization of the DiI-labelled cells arising in the MGE showed that those which migrated into the marginal zone and showed features of Cajal-Retzius cells also expressed reelin. This secreted protein, a feature of Cajal-Retzius cells in the developing cerebral cortex and hippocampus, appears to be crucial for the establishment of normal lamination in the cortical plate (Ogawa et al 1995, Frotscher 1998). However, we observed that these cells did not express calretinin, a calcium-binding protein often used as a marker of Cajal-Retzius cells

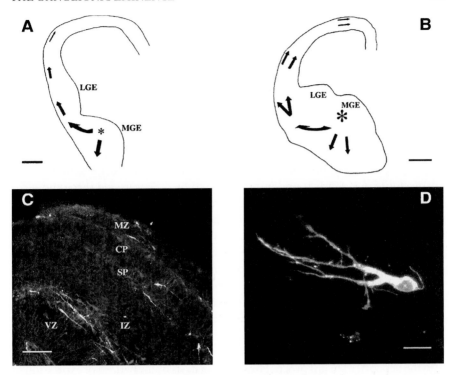

FIG. 3.   (A, B) Camera lucida drawings of coronal sections through part of the rat forebrain following placement of DiI (1,1'dioctadecyl-3,3,3',3'-tetramethylindocarbocyanine perchlorate) crystals in the medial ganglionic eminence (MGE; asterisks) at embryonic day (E) 14 (A) and E16 (B). Arrows indicate the direction of migration of labelled cells in the cortical primordium two (A) and three (B) days after DiI application. In A, labelled cells appeared as a stream rounding the corticostriatal sulcus and heading towards the preplate. In B, migrating cells were directed predominantly towards the most superficial and deeper aspects of the cortical primordium. The disposition of a number of DiI-labelled cells in these layers is shown in (C). (D) Labelled migrating neuron in the intermediate zone (IZ) three days after application of DiI in the MGE. CP, cortical plate; LGE, lateral ganglionic eminence; MZ, marginal zone; SP, subplate; VZ, ventricular zone. Bars = 350 μm, 600 μm, 100 μm and 20 μm for A, B, C and D, respectively (from Lavdas et al 1999).

(Del Río et al 1995), supporting the notion that cells in the marginal zone comprise a heterogeneous group of neurons. This was initially postulated for primates including humans (Meyer & Goffinet 1998, Supèr et al 1998), but work in other species has also shown that cells in the marginal zone show diverse morphologies, and complex and different neurochemical profiles and fates (Bradford et al 1977, Parnavelas & Edmunds 1983, Meyer et al 1998). One of the groups of neurons populating the marginal zone that has received attention since the early part of the century are the so-called subpial granule neurons. Recent work by Meyer et al

(1998) has indicated that such cells, originating in a restricted sector of the telencephalic vesicle, also exist in the rat cortex. These authors further suggested that the derivatives of the subpial granule cells migrate into the superficial part of the marginal zone and differentiate into Cajal-Retzius cells. It may be that these neurons correspond to the Cajal-Retzius cells described here and have their origin in the MGE (Lavdas et al 1999).

The origin, migration and distribution of these early neurons coincided with the pattern of expression of the novel LIM homeobox gene *Lhx6* in the developing telencephalon (Grigoriou et al 1998) and, indeed, we found that DiI-labelled MGE cells expressed this transcription factor. In the later stages of neurogenesis and in postnatal life, *Lhx6* was found to be expressed by cells scattered throughout the cortical thickness. Although double-labelling studies have not yet been conducted, their distribution in the cortex resembled that of GABAergic neurons, suggesting that GABA-containing neurons in the early stages of cortical development and in later life express *Lhx6*. This raises the possibility that the expression of GABA is under the control of this LIM homeobox gene. Furthermore, differential expression of *Lhx6* in a migratory population of MGE cells suggests that products of this gene, uniquely or in combination with other transcription factors, might play a role in the decision of MGE cells to differentiate *in situ* or migrate to the neocortex.

The molecular mechanisms that control the migration of cells from the MGE into the neocortex are unknown. However, analysis of the pattern of expression of *Dlx1* and *Dlx2* in the MGE (Bulfone et al 1993, Grigoriou et al 1998) suggests that the dorsal migration of MGE-derived cells may also be under the control of this subfamily of homeobox genes. This hypothesis is further supported by the absence of *Lhx6*-expressing cells in the cortex of *Dlx1/2* null mouse embryos (S. A. Anderson & J. L. R. Rubenstein, unpublished results 1999). Genetic evidence that the MGE is an important source of cortical neurons also comes from analysis of mice with a mutation in the *Nkx2.1* homeobox gene. In the absence of *Nkx2.1*, pallidal regions of the telencephalon do not form (e.g. the MGE), and they take on characteristics of the more dorsal regions (e.g. the LGE; Sussel et al 1999). The mice also have a severe reduction in the number of cortical interneurons. In addition, *Mash1* mutants, which have a greatly reduced MGE, also show reduced numbers of cortical interneurons (Casarosa et al 1999). However, because the LGE is also defective in these mice, it is difficult to determine whether the cortical interneuron deficiency is due to abnormalities in the MGE or LGE.

Analysis of mutants with differing basal telencephalic abnormalities provides the opportunity for determining the relative contributions of the LGE and MGE to the cortex and olfactory bulb (Anderson et al 1999). For example, the *Nkx2.1* mutants that appear to lack a functional MGE show roughly a 50% reduction in GABA-positive cells in the neocortex (at E18.5; Sussel et al 1999). However, the olfactory bulbs of these mutants contain nearly a normal number of GABA-positive cells. In slice cultures of these mutants, a greatly reduced number of cells are seen migrating into the neocortex following placement of DiI crystals in either the LGE or in the position of the abnormal 'MGE' (S. A. Anderson & J. L. R. Rubenstein, unpublished results 1999). In contrast, the *Dlx1/2* mutant, which lacks migration out of both the LGE and the MGE in slice culture (Anderson et al 1997a, Anderson & Rubenstein 2000), has a roughly 75% reduction of GABA cells in the neocortex and virtually no GABA-positive cells in the olfactory bulb. These results suggest that the MGE is the source for many neocortical interneurons, while the LGE provides some interneurons to neocortex and many to the olfactory bulb (Anderson et al 1999). This scenario is supported by the finding that dissociated MGE cells injected into the lateral ventricle mainly migrate into the pallidum, striatum and neocortex, while LGE cells migrate predominantly into the striatum and olfactory bulb (Wichterle et al 1999).

References

Anderson SA, Eisenstat DD, Shi L, Rubenstein JLR 1997a Interneuron migration from basal forebrain to neocortex: dependence on *Dlx* genes. Science 278:474–476

Anderson SA, Qiu M, Bulfone A et al 1997b Mutations of the homeobox genes *Dlx-1* and *Dlx-2* disrupt the striatal subventricular zone and differentiation of late born striatal neurons. Neuron 19:27–37

Anderson SA, Mione M, Yun K, Rubenstein JLR 1999 Differential origins of neocortical projection and local circuit neurons: role of *Dlx* genes in neocortical interneurogenesis. Cereb Cortex 9:646–654

Bradford R, Parnavelas JG, Lieberman AR 1977 Neurons in layer I of the developing occipital cortex of the rat. J Comp Neurol 176:121–132

Bulfone A, Puelles L, Porteus MH, Frohman MA, Martin GR, Rubenstein JLR 1993 Spatially restricted expression of *Dlx-1*, *Dlx-2* (*Tes-1*), *Gbx-2*, and *Wnt-3* in the embryonic day 12.5 mouse forebrain defines potential transverse and longitudinal segmental boundaries. J Neurosci 13:3155–3172

Casarosa S, Fode C, Guillemot F 1999 *Mash1* regulates neurogenesis in the ventral telencephalon. Development 126:525–534

De Carlos JA, López-Mascaraque L, Valverde F 1996 Dynamics of cell migration from the lateral ganglionic eminence in the rat. J Neurosci 16:6146–6156

Del Río JA, Martínez A, Fonseca M, Auladell C, Soriano E 1995 Glutamate-like immunoreactivity and fate of Cajal-Retzius cells in the murine cortex as identified with calretinin antibody. Cereb Cortex 5:13–21

Derer P, Derer M 1990 Cajal-Retzius cell ontogenesis and death in mouse brain visualized with horseradish peroxidase and electron microscopy. Neuroscience 36:839–856

Frotscher M 1998 Cajal-Retzius cells, Reelin, and the formation of layers. Curr Opin Neurobiol 8:570–575

Gray GE, Sanes JR 1991 Migratory paths and phenotypic choices of clonally related cells in the avian optic tectum. Neuron 6:211–225

Grigoriou M, Tucker AS, Sharpe PT, Pachnis V 1998 Expression and regulation of *Lhx6* and *Lhx7*, a novel subfamily of LIM homeodomain encoding genes, suggests a role in mammalian head development. Development 125:2063–2074

Lavdas AA, Grigoriou M, Pachnis V, Parnavelas JG 1999 The medial ganglionic eminence gives rise to a population of early neurons in the developing cerebral cortex. J Neurosci 19:7881–7888

Luskin MB, Parnavelas JG, Barfield JA 1993 Neurons, astrocytes, and oligodendrocytes of the rat cerebral cortex originate from separate progenitor cells: an ultrastructural analysis of clonally related cells. J Neurosci 13:1730–1750

Marín-Padilla M 1998 Cajal-Retzius cells and the development of the neocortex. Trends Neurosci 21:64–71

McConnell SK, Ghosh A, Shatz CJ 1994 Subplate pioneers and the formation of descending connections from cerebral cortex. J Neurosci 14:1892–1907

Meyer G, Goffinet AM 1998 Prenatal development of reelin-immunoreactive neurons in the human neocortex. J Comp Neurol 397:29–40

Meyer G, Soria JM, Martínez-Galán JR, Martín-Clemente B, Fairén A 1998 Different origins and developmental histories of transient neurons in the marginal zone of the fetal and neonatal rat cortex. J Comp Neurol 397:493–518

Mione MC, Danevic C, Boardman P, Harris B, Parnavelas JG 1994 Lineage analysis reveals neurotransmitter (GABA and glutamate) but not calcium-binding protein heterogeneity in clonally related cortical neurons. J Neurosci 14:107–123

Mione MC, Cavanagh JFR, Harris B, Parnavelas JG 1997 Cell fate specification and symmetrical/asymmetrical divisions in the developing cerebral cortex. J Neurosci 17:2018–2029

Ogawa M, Miyata T, Nakajima K et al 1995 The *reeler* gene-associated antigen on Cajal-Retzius neurons is a crucial molecule for laminar organization of cortical neurons. Neuron 14:899–912

O'Rourke NA, Sullivan DP, Kaznowski CE, Jacobs AA, McConnell SK 1995 Tangential migration of neurons in the developing cerebral cortex. Development 121:2165–2176

Parnavelas JG, Edmunds SM 1983 Further evidence that Retzius-Cajal cells transform to non-pyramidal neurons in the developing rat visual cortex. J Neurocytol 265: 863–871

Parnavelas JG, Dinopoulos A, Davies SW 1989 The central visual pathways. In: Björklund A, Hökfelt T, Swanson LW (eds) Handbook of chemical neuroanatomy, vol 7: Integrated systems of the CNS, Part II. Elsevier, Amsterdam, p 1–164

Parnavelas JG, Barfield JA, Franke E, Luskin MB 1991 Separate progenitor cells give rise to pyramidal and non-pyramidal neurons in the rat telencephalon. Cereb Cortex 1:463–468

Porteus MH, Bulfone A, Liu JK, Puelles L, Lo LC, Rubenstein JLR 1994 DLX-2, MASH-1, and MAP-2 expression and bromodeoxyuridine incorporation define molecularly distinct cell populations in the embryonic mouse forebrain. J Neurosci 44:6370–6383

Rakic P 1988 Specification of cerebral cortical areas. Science 241:170–176

Reid CB, Liang I, Walsh C 1995 Systematic widespread clonal organization in cerebral cortex. Neuron 15:299–310

Rockel AJ, Hiorns RW, Powell TP 1980 The basic uniformity in structure of the neocortex. Brain 103:221–244

Soriano E, Dumesnil N, Auladell C, Cohen Tannoudji M, Sotelo C 1995 Molecular heterogeneity of progenitors and radial migration in the developing cerebral cortex revealed by transgene expression. Proc Natl Acad Sci USA 92:11676–11680

Supèr H, Soriano E, Uylings HBM 1998 The functions of the preplate in development and evolution of the neocortex and hippocampus. Brain Res Rev 27:40–64

Sussel L, Marin O, Kimura S, Rubenstein JLR 1999 Loss of *Nkx2.1* homeobox gene function results in a ventral to dorsal molecular respecification within the basal telencephalon: evidence for a transformation of the pallidum into the striatum. Development 126:3359–3370

Tamamaki N, Fujimori KE, Takauji R 1997 Origin and route of tangentially migrating neurons in the developing neocortical intermediate zone. J Neurosci 17:8313–8323

Tan SS, Kalloniatis M, Sturm K, Tam PPL, Reese BE, Faulkner-Jones B 1998 Separate progenitors for radial and tangential cell dispersion during development of the cerebral cortex. Neuron 21:295–304

Uylings HBM, van Eden CG, Parnavelas JG, Kalsbeek A 1990 The prenatal and postnatal development of the rat cerebral cortex. In: Kolb E, Tees RC (eds) The cerebral cortex of the rat. MIT Press, Cambridge, MA, p 35–76

Walsh C, Cepko CL 1992 Widespread dispersion of neuronal clones across functional regions of the cerebral cortex. Science 255:434–440

Walsh C, Cepko CL 1993 Clonal dispersion in proliferative layers of developing cerebral cortex. Nature 362:632–635

Wichterle H, Garcia-Verdugo JM, Herrera DG, Alvarez-Buylla A 1999 Young neurons from medial ganglionic eminence disperse in adult and embryonic brain. Nat Neurosci 2:461–466

DISCUSSION

Reiner: Do Lhx6-positive cells populate the claustrum and the amygdala?

Parnavelas: We haven't done extensive studies with the anti-Lhx6 antibody; we have only focused on the cortex, but we will soon investigate other areas.

Welker: In order to get from the medial ganglionic eminence (MGE) to the cerebral cortex, the cells have to pass through the lateral ganglionic eminence (LGE), so is it possible that cells coming from the MGE are responsible for the labelling you observe in the LGE? Secondly, do you know the birth dates of your migrating cells?

Parnavelas: Some cells emanating in the MGE seem to go through the LGE on their way to the cortex; others seem to go around it. We do not know the birth dates of these cells or the order in which they are deposited.

Welker: But the cells interact at the level of the cortical plate, so cells of the same age move together towards a common destination layer. For example, pyramidal cells of the same age move from the ventricular zone to layer II–III where they intermingle with the population of γ-aminobutyric acid (GABA)ergic interneurons.

Parnavelas: Thymidine autoradiography studies in the rat have shown that the inside-out pattern of cortical cell generation does not apply as strictly to non-pyramidal cells as it does to pyramidal neurons. These cells tend to be more diffusely positioned in the cortex than their pyramidal counterparts (Cavanagh & Parnavelas 1988).

Welker: My last question concerns spiny stellate cells. Do you classify these as pyramidal cells?

Parnavelas: There are only few spiny stellate cells in the rat cerebral cortex; they are found mainly in layer IV.

Karten: This is a point that concerns me also. The layer IV cells that are the target of the specific thalamic projections are small cells. They are glutaminergic, non-GABAergic cells, and we don't traditionally classify them as being pyramidal cells. We traditionally characterize them as being small stellate cells. This is typical of the striatum, the auditory cortex and the barrel fields of the cortex.

Parnavelas: Who has shown that the cells in layer IV which receive thalamic input are glutaminergic, non-GABAergic cells?

Welker: There are EM studies which showed that spiny stellate cells in layer IV of the mouse barrel cortex receive direct thalamic input. Interestingly, the distribution of these thalamic synapses on spiny stellate cells differ from those on GABAergic interneurons (White 1989).

Parnavelas: This is correct. However, we were not able to distinguish between spiny and smooth stellate cells in our studies.

O'Leary: At least in rodent barrel cortex, a substantial proportion of cells in layer IV are spiny stellate cells. However, recent studies have shown that early in development, layer IV stellate cells have a morphology that resembles small pyramidal cells. They have an apical dendrite that extends to layer I; they gradually retract this dendrite and elaborate additional ones as they take on the stellate morphology.

Parnavelas: None of the cells I have described here appeared pyramidal in form; they were also found to contain GABA. They are, by definition, cortical interneurons.

O'Leary: If one uses only morphology to assess whether a neuron is pyramidal or non-pyramidal within a given clone, then given their similarities early in their development, the stage at which a clonal analysis is done may have a large impact on the interpretation.

Parnavelas: Of course the time at which clonal analysis is done is crucial. Lineage studies in adult rats have suggested that each cortical progenitor cell gives rise to progeny of the same phenotype (homogeneous clones). However, similar studies during development have shown that neuronal clones display morphological and neurotransmitter heterogeneity (Lavdas et al 1996).

Broccoli: I can confirm that there is heterogeneity among Cajal-Retzius cells. In fact, at least two different populations of these cells have been recently identified (Meyer et al 1998): the first is generated in the ventricular zone starting from E10 and migrates radially; whereas the second differentiated two days later from the retrobulbar ventricle and follows a subpial tangential migration. Moreover, in mice homozygous for an *Emx2* null allele it has been observed that while the first-born Cajal-Retzius neurons develop normally, the later differentiated neurons disappear, leading to neural migration and layer stratification defects in *Emx2* mutant brains. The second class of these neurons seems to be *Emx2*-dependent (Mallamaci et al 2000). I would like to ask to John Rubenstein if he has seen any abnormalities in the development of the cerebral cortex in *Nkx2.1* knockouts, that are missing the ventral basal ganglia.

Rubenstein: We haven't yet looked at the expression of reelin, but we have looked at the expression of calretinin in the marginal zone, and found that it was unchanged compared to wild-type animals. We haven't yet done birth dating, but using gene markers and Nissl staining, we found that there were no obvious problems with regard to lamination or regionalization of any cortical area. The only aberration was that all the calbindin staining disappeared in the lateral cortex, and possibly also in the claustrum, neocortex and hippocampus.

Goffinet: How well accepted is the suggestion that Cajal-Retzius cells have multiple origins?

Parnavelas: There is no doubt that there are different origins and different types of marginal zone cells. Some are the unique Cajal-Retzius cells. However, the term 'Cajal-Retzius' is used rather loosely nowadays, encompassing not only Cajal-Retzius cells, but a variety of other neuronal cell types.

Goffinet: I have another comment. It's clear from chimera data which the Mullen and Mikoshiba groups published a long time ago that only a low level of reelin is required for the development of a normal cortex (Mullen 1977, Mikoshiba et al 1985). About 20% of wild-type levels is probably enough. Therefore, one can imagine that if some Cajal-Retzius precursors are knocked out, enough reelin would still be present for the development of a normal cortical plate.

Karten: We have raised some fundamental aspects regarding cortical development. First of all, there is the issue of tangential migration, which has been a subject of considerable dispute. There are now a number of lines of evidence suggesting that cells which contribute to the formation of the cortex — Cajal-Retzius, interneurons and others — don't necessarily arrive from the pallial ependyma. This raises the issue of what's going on in the pallial ependyma and whether the general notion that neurons migrate radially from the pallial ependyma is correct.

Another issue is whether all mitotic activity occurs only within the ventricular zone, or whether there are ectopic zones. It has been suggested that there is an area deep in the MGE which has an independent zone of mitotic activity, although this suggestion has never been properly documented. It's not clear to me from these various claims whether everything can be simply put into the ventricular zone as a to-and-fro migration/mitotic activity.

Broccoli: Is tangential migration conserved in evolution?

Puelles: In collaboration with John Rubenstein's and Salvador Martinez's groups, we have done fate mapping of chicken LGE and MGE primordia. We found substantial evidence for tangential migration from the MGE into the whole pallium (Cobos et al 2000), but scarce, if at all, from the LGE. These migrated cells expressed calbindin and many of them seem to be GABAergic. Moreover, *Dlx*- and GABA-expressing neurons separately have been found in the frog pallium (Dirksen et al 1993, Barale et al 1996), which allows the possibility that subpallial cells also migrate into the pallium in anura, though this has not been proven. This means that the migration we are speaking about may have originated at least in stem amniotes, if not earlier.

Boncinelli: Can you estimate the birth date of these cells?

Puelles: Not without specific experiments. However, the MGE starts to differentiate rather early (three to four days *in ovo*).

Karten: Let's focus on the concept that we've believed in for 100 years, i.e. cortical neurons arise from pallial ependyma. We are now at the stage where almost all of us are saying that the situation is much more complicated than this.

Rakic: I would like to suggest that the ganglionic eminence may be a specialization of the subventricular zone. The ganglionic eminence becomes enlarged because there is increased proliferative activity, but that doesn't necessarily mean that it is fundamentally different.

Karten: This may just be a question of semantics. Ian Smart suggested in the 1960s that there were serious problems with how we view tangential migration. The pallial ependyma as a source for cortical neurons has been so rich a source for experiments that we have tended to focus on it. But perhaps we have now reached the stage where we have to look at more specific details. It's beginning to look as if we just can't say 'this is pallial'. The length of the pathway and the guidance you would have to postulate for this is just a little too difficult.

Puelles: My feeling is that a discussion of semantics is not useful for us now. We have enough novel genetic evidence to help us distinguish between pallial and non-pallial patterns of pre-specification, which are quite distinct. Of course, everything (pallium and subpallium) is telencephalic, but that does not help so much as keeping the differential characters in mind, with one eye on the complexities introduced by tangential migrations. There are unexpected patterns, i.e. many cells destined for the cortex come from the subpallium, but it is also true that some cells from the pallium migrate to the subpallium. For instance, Striedter et al (1998) showed by *in ovo* dextranamine labelling in chick embryos that some cells in the olfactory tuberculum, a subpial part of the subpallium, come from the pallium. The existence of such cell movements means we have to partially change the meaning of 'pallial' and 'subpallial' again (distinguish 'primary pallial' cells from 'secondary pallial' ones; similar for subpallium), but the distinction of what is primarily pallial or subpallial is clear at the moment.

O'Leary: What is the current opinion of how much tangential migration occurs within the neocortex of cells generated from the neocortical neuroepithelium?

Parnavelas: There is a substantial amount of tangential migration within the cortex of cells generated from the cortical neuroepithelium. We are attempting to obtain a quantitative estimate by using green fluorescence protein-expressing retroviruses.

O'Leary: I also have a follow-up question for John Rubenstein. He showed us some sharp gene expression borders, and I would like to know his thoughts on how these sharp expression borders may come about if there is so much tangential migration of cortical neurons.

Rubenstein: It could be controlled by rapid degradation of RNA and protein, and tight control of transcription.

Levitt: But John Parnavelas said that there are 70–75% glutaminergic and 25–30% GABAergic neurons in the cortex. The *in situ* hybridization studies of transcription factors indicate that there are sharp boundaries and an over-abundance of projection neurons compared with interneurons. The estimates of tangential migration are about 25–30%, which suggests that the tangentially migrating neurons are the interneurons and not the projection neurons. I don't know of anyone who has done a serial reconstruction and counted how many are double-labelled. Do all the cells in a particular cortical area express a transcription factor, or only 70–80%? Tan et al (1998) have shown, using glutamate staining, that there is virtually no spread of projection neurons, which suggests that at least 70% of neurons are migrating radially.

In the monkey, the thymidine injection results suggest that there are sharp boundaries in terms of laminar distribution; it is incredibly sharp. Therefore, if tangential migration of GABAergic neurons from the subpallium occurs in the primate, how do the interneurons from the subpallial region migrate to the pallial region, and subsequently assemble temporarily and spatially in a way that's linked to the sharp laminar distribution of neurons born on a particular day?

Parnavelas: It is possible that there is a mechanism in the primate which can account for this.

Rakic: In monkeys, the distinction between layers is sharp. Isochronically labelled cells form in a single, well-defined line. When we first discovered this, we did not envisage that a separate mechanism was involved, but rather that it could be explained by the longer development period (60 days as opposed to five days in the rat), which provide higher temporal resolution. On the other hand, it is possible that primate cortex is more precisely organized.

Karten: One of the reasons that so many people have focused on the striate (visual) cortex is because it is such a distinct area relative to the rest of the pallium. However, the temporal cortex develops before the medial cortex. This raises a different question, i.e. the temporal lobe may be developing by a some-what different mechanism. We are assuming that all cortex arises by a singular mechanism and by a common pattern of migration, but this may not be the case.

Puelles: What is the evidence for postulating that the temporal cortex is different?

Karten: I'm just raising this as a possibility.

Rubenstein: The central nervous system, whether in the spinal cord or in the telencephalon, makes different kinds of cell types with different dorsoventral positions using morphogen gradients. Then, depending on the type of cells, they have different programmes that determine whether they migrate radially to become motor neurons, secondary sensory neurons or cortical pyramidal cells, or whether they migrate tangentially and become interneurons and oligo-dendrocytes. I believe that this is a general phenomenon everywhere in the central nervous system. Our evidence suggests that in the telencephalon, interneurons migrate to all pallial regions, i.e. the hippocampus, piriform cortex, olfactory bulb and neocortex.

Bonhoeffer: What is the nature of migration? Does the growth cone pull the cell, or is the growth cone pushed by the cell body?

Parnavelas: In the majority of cells the growth cone leads the migration, which suggests that it is pulling the cell.

Bonhoeffer: Is this a reasonable way to guide the cell? Because if there are various growth cones pulling in different directions, they may become stuck with other connections. Perhaps the other connections are not present.

Karten: You also have to postulate differential trafficking within the cell, so that the nucleus knows which of the processes it has to follow.

Puelles: We find that in general migrating cells only have a single leading process.

Parnavelas: We also found that the majority of migrating cells only had a single process with the growth zone at the tip.

Karten: It is possible that the cells you're studying only have one process, but Morest's data (Zhou et al 1996) and Rubel's data (Oesterle et al 1997) indicate that cells continue to migrate as they are spreading out their dendrites. This was one of the early points to come out of the signalling studies.

Bonhoeffer: Do you know anything about the gradients that guide these cells or the gradients that cause them to stop?

Parnavelas: No. This is an area we wish to explore.

O'Leary: In collaboration with Marc Tessier-Lavigne, we have been looking at directed cell migration in the hindbrain. We have found that axon guidance molecules — e.g. netrin-1 which helps guides commissural axons to the ventral midline in the hindbrain and spinal cord — are also involved in directing non-radial, long-distance neuronal migration in the hindbrain.

I also have question for both John Parnavelas and John Rubenstein. It seems as if the timing and location of the cell migration that you both observe arising from the ganglionic eminence, at least for those cells found in the intermediate zone, is consistent with the notion that they may be using the efferent pathway as a guidance system to navigate them to and through the cortex. This may be similar to the handshake hypothesis. Therefore, I wonder whether there are fewer migratory cells in the *Gbx2* mutant, which lacks a thalamocortical projection and has a perturbed cortical output projection.

Rubenstein: We have not yet done interneuron staining in out *Gbx2* mutant. We have done interneuron staining in the *Tbr* mutant, which also lacks thalamic inputs into the cortex, and we find that the interneurons are still present.

Parnavelas: I have a suspicion that there are multiple mechanisms involved in this process. Use of the efferent pathway may be one of the mechanisms.

Karten: It seems as though there is also a glial-to-glial transfer of migrating neurons, just like a transfer ticket on a bus, which raises the question of whether the cells can make the wrong transfers in the same way as it is possible to get on the wrong bus, and what is the nature of the ticket?

Molnár: The early corticofugal projections which pause at the intermediate zone–striatal junction were suggested to be responsible for the guidance of these cells. We were thinking of doing similar experiments in *reeler* mice to see if the migrating cells jump onto what they think is the correct bus but then end up in the wrong place in the superplate.

Pettigrew: Do we know that the tangentially migrating neurons know where they are going? Could they just be filing in one-by-one and filling up the space?

Rubenstein: They do follow particular pathways through the marginal zone and intermediate zone. They don't just plough straight into the cortical plate.

Karten: But Jack Pettigrew is raising a quantitative issue, i.e. perhaps the cells keep migrating until the space is filled.

Herrup: This is also relevant in the cerebellum. How do the external granule cells know how to fill the surface of the cerebellum, and how do they know when to stop? I can't answer this.

Pettigrew: Is there any evidence that the tangentially migrating interneurons know where to go tangentially? I can imagine that there will be specific signals that determine which layer they go to, but how do they know where to go tangentially?

Puelles: In the chick fate-mapping experiments I mentioned before (Cobos et al 2000), there are cases with small grafts, where only small portions of the MGE become labelled. The cells that migrate out of these patches go everywhere in the overlying pallium (both dorsal ventricular ridge [DVR] and cortex, i.e. Wulst or hippocampus). There doesn't seem to be specific migration targets in this system.

Reiner: The chick DVR doesn't have the same migration path cues as there are in the neocortex.

Rakic: Jack Pettigrew's question is interesting, and the intuitive answer would be that there is no specificity because they disperse widely across the cortex, but this doesn't mean that they don't have specific pathways. One can imagine that pyramidal cells migrate radially from precise connections within subcortical structures, whereas stellate cells could be performing the same function in many different areas. The latter cells would then be less specific than pyramidal cells that form precise point-to-point connectivity. Stellate cells may nevertheless use specific pathways as a substrate to carry them to the proper position, although individual cells could end up in one or the other cortical area and perform the same function.

Herrup: If this were a question of axon guidance, I expect we would take a different view. We would be horrified to think that axons just filled up the space, because we think they care about targeting. We've been discussing that migratory mechanisms might be similar to axon elongation mechanisms, and as a default position I believe we ought to assume that the cells know where they are going.

Boncinelli: For how long do these cells migrate? Do they still migrate after birth?

Parnavelas: In rats, it appears that the last tangentially migrating neurons leave the ganglionic eminence around embryonic day (E) 17.

Levitt: van der Kooy and colleagues did cell aggregation studies to look at simple neighbour relationships that develop (Krushel et al 1995). They did striatal–striatal, cortical–cortical and striatal–cortical aggregates. By nearest-neighbour analysis they found that the cells which associate with each other ignore site of origin. Rather, they sort mostly according to their time of origin.

Lumsden: We have evidence that striatal and pallial cells segregate from each other during a particular transient period in development, at around E14 in the rat; but by E18 the cells no longer segregate. We found that chick striatal and dorsal telencephalic cells also segregate in these aggregation assays but, interestingly, we found that cells from the intervening region, of the DVR, mix happily with both the striatal and dorsal telencephalic cells. In all cases, the ability of young cells to segregate according to region of origin seems to rely on calcium-dependent adhesion molecules. We think that this selective adhesion phenomenon might help to restrict cell mixing and maintain positional information during forebrain development (Götz et al 1996).

References

Barale E, Fasolo A, Girardi E, Artero C, Franzoni MF 1996 Immunohistochemical investigation of γ aminobutyric acid ontogeny and transient expression in the central nervous system of *Xenopus laevis* tadpoles. J Comp Neurol 368:285–294

Cavanagh ME, Parnavelas JG 1988 Neurotransmitter differentiation in cortical neurons. In: Parnavelas JG, Stern CD, Stirling RV (eds) The making of the nervous system. Oxford University Press, Oxford, p 434–453

Cobos I, Puelles L, Rubenstein JLR, Martinez S 2000 Fate map of the chicken rostral neural plate at stages 7–8, using quail chick homotopic grafts. in prep

Dirksen ML, Mathers P, Jamrich M 1993 Expression of a *Xenopus Distal-less* homeobox gene involved in forebrain and cranio-facial development. Mech Dev 41:121–128

Götz M, Wizenmann A, Reinhardt S, Lumsden A, Price J 1996 Selective adhesion of cells from different telencephalic regions. Neuron 16:551–564

Krushel LA, Fishell G, van der Kooy D 1995 Pattern formation in the mammalian forebrain: striatal patch and matrix neurons intermix prior to compartment formation. Eur J Neurosci 7:1210–1219

Lavdas AA, Mione MC, Parnavelas JG 1996 Neuronal clones in the cerebral cortex show morphological and neurotransmitter heterogeneity during development. Cereb Cortex 6:490–497

Mallamaci A, Mercurio S, Muzio L, Cecchi C, Pardini C, Boncinelli E 2000 The lack of *Emx2* causes impairment of Reelin signaling and defects of neuronal migration in the developing cerebral cortex. J Neurosci 20:1109–1118

Meyer G, Soria JM, Martinez-Galan JR, Martin-Clemente B, Fairen A 1998 Different origins and developmental histories of transient neurons in the marginal zone of the fetal and neonatal rat cortex. J Comp Neurol 397:493–518

Mikoshiba K, Yokohama M, Terashima T et al 1985 Analysis of abnormality of corticohistogenesis in the *reeler* mutant mice by producing chimera mice: absence of abnormality of neurons in the reeler. Acta Histochem Cytochem 18:113–124

Mullen RJ 1977 Genetic dissection of the CNS with mutantnormal mouse and rat chimeras. Neurosci Symp 2:47–65

Oesterle EC, Tsue TT, Rubel EW 1997 Induction of cell proliferation in avian inner ear sensory epithelia by insulin-like growth factor and insulin. J Comp Neurol 380:262–274

Striedter GF, Marchant TA, Beydler S 1998 The 'neostriatum' develops as part of the lateral pallium in birds. J Neurosci 18:5839–5849

Tan SS, Kalloniatis M, Sturm K, Tam PP, Reese BE, Faulkner-Jones B 1998 Separate progenitors for radial and tangential cell dispersion during development of the cerebral neocortex. Neuron 21:295–304

White EL 1989 Cortical circuits: synaptic organization of the cerebral cortex. Birkhäuser, Basel

Zhou X, Hossain WA, Rutledge A, Baier C, Morest DK 1996 Basic fibroblast growth factor (FGF-2) affects development of acoustico-vestibular neurons in the chick brain *in vitro*. Hear Res 101:187–207

Conserved developmental algorithms during thalamocortical circuit formation in mammals and reptiles

Zoltán Molnár[1]

Institut de Biologie Cellulaire et de Morphologie, Université de Lausanne, Rue du Bugnon 9, 1005 Lausanne, Switzerland

Abstract. The general patterns of early thalamocortical development follow a similar sequence in all mammals. Thalamocortical projections descend through the ventral thalamus, advance in the internal capsule amongst cells which already possess dorsal thalamic projections, then reach the cerebral cortex by associating with subplate cells and their early corticofugal projections. Initially, the thalamic projections pause in the internal capsule and subplate layer. The interactions of the thalamocortical projections with the early generated, largely transient cells of the subplate, marginal zone, internal capsule and ventral thalamus are believed to play a crucial role in the organized deployment of thalamic projections and establishing a functional cortical architecture. Selective fasciculation, contact guidance and release of neurotrophic factors are thought to play roles in the development of thalamocortical projections. These ideas are obtaining support from recent work on *reeler* and other strains of mice. The evolutionary origin of these largely transient cells and the overlying logic of early developmental steps are not understood. The behaviour of the thalamocortical and corticothalamic projections at the corticostriatal junction is particularly puzzling. The comparison of early forebrain development in mammals and reptiles is beginning to reveal highly conserved cellular and molecular interactions during early thalamocortical development and to reveal homologies between telencephalic subdivisions.

2000 Evolutionary developmental biology of the cerebral cortex. Wiley, Chichester (Novartis Foundation Symposium 228) p 148–172

Due to advances in embryonic tracing and imaging techniques, organotypic culture systems and analysis of mutant and transgenic mice, the understanding of the cellular and molecular aspects of forebrain development is rapidly increasing. Carbocyanine dye tracing has revealed the degree of order that the thalamic fibres maintain, and the cellular elements they encounter while they grow out from the diencephalon, through the internal capsule and accumulate below the

---

[1] Current address: Department of Human Anatomy and Genetics, University of Oxford, South Parks Road, Oxford OX1 3QX, UK.

corresponding cortical region in various species (McConnell et al 1989, Catalano et al 1991, reviewed in Allendoerfer & Shatz 1994, Molnár 1998). Thalamic afferents reach the cortex by associating with pre-existing cells and preplate projections (Erzurumlu & Jhaveri 1992, De Carlos & O'Leary 1992, Molnár & Blakemore 1995, Molnár et al 1998a,b,c). These guiding mechanisms are thought to be important for the thalamic fibres to find their way to the cortex through the subdivisions of the embryonic forebrain.

The handshake hypothesis proposes that axons from the thalamus and from the early-born cortical preplate cells meet and intermingle in the basal telencephalon, so that thalamic axons grow over the scaffold of preplate axons and become 'captured' for the waiting period in the subplate (Blakemore & Molnár 1990, Molnár & Blakemore 1990, 1995). Detailed tracing studies in the rat, mouse and marsupials summarized in Fig. 1A show that early corticofugal projections are in a position to assist thalamocortical fibres through the distal part of their journey towards the cortex:

(1) Cortical efferents grow through the intermediate zone and descend into the internal capsule in an organized fashion (Molnár et al 1998a).

(2) Thalamic afferents show regional specificity in their targeting from the time they arrive in the subplate, prior to entry in the cortical plate (Blakemore & Molnár 1990, Catalano et al 1996).

(3) Double labelling studies combined with 3D confocal microscopic reconstructions demonstrate close contact between the early corticofugal and developing thalamocortical projections in the internal capsule and intermediate zone as they both arrive and travel through the region (Molnár et al 1998a,b).

(4) Labelling from a single cortical carbocyanine crystal placement, after thalamic fibres arrive in the cortex, was found to anterogradely label corticofugal and retrogradely label thalamocortical fibres which together form a single bundle, suggesting that the two fibre systems lie in close proximity and follow similar trajectories (Blakemore & Molnár 1990, Molnár et al 1998a).

However, these tracing studies support a possible role of these interactions in guidance, they do not prove that this guiding mechanism is indeed essential for thalamocortical development.

Distinct fasciculation patterns of thalamocortical projections and their altered development in L1 knockout, *reeler* mutant and other transgenic mice

Thalamocortical axons show distinct patterns of organization along their path through telencephalic subdivisions to the cortex. As they pass through the

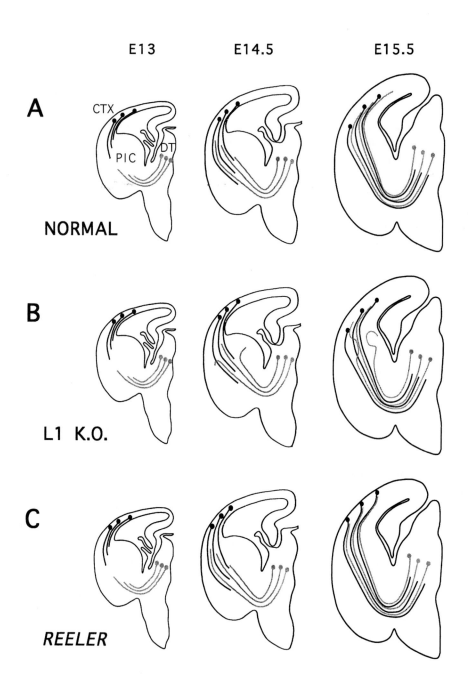

primitive internal capsule and encounter the anlage of the corpus striatum, the array of thalamocortical fibres form distinct fascicles, which open up in a fan-shaped fashion. At the border between the corpus striatum and the intermediate zone the thalamic fibres defasciculate and reassemble into a fairly uniform array of individual fibres (Molnár et al 1998a, Fig. 3C). The trajectories of the axons remain parallel as they turn into the intermediate zone. The molecular nature of the fasciculation and the properties of the extracellular environment are not yet fully understood. Subplate neurons are known to express immunoreactivity to the surface molecules L1 (Godfraind et al 1988, Chung et al 1991), fibronectin (Stewart & Pearlman 1987) glycosaminoglycans (Derer & Nakanishi 1983) and chondroitin sulphate core proteins (Miller et al 1995, Bicknese et al 1994). These could provide highly attractive substrates for the growth of thalamic axons expressing appropriate adhesion molecules in an otherwise relatively non-permissive environment. It is yet to be determined whether thalamic fibres require more than one of these extracellular cues (sequentially or simultaneously) for their guidance or whether there is a specific requirement for only one or a subset of them. However, one potential guidance cue may be provided by neurocan,

---

FIG. 1.   Schematic summary diagram on the early development of the reciprocal thalamocortical and corticofugal projections in wild-type (A), L1 knockout mouse (B) and *reeler* mutant mouse (C) at different embryonic stages (E13, E14.5, E15.5). Each diagram represents an imaginary section through the right hemisphere, revealing the entire pathway. (A) In normal mice, at E13, the early thalamic and corticofugal fibres synchronously approach the internal capsule where they pause until E14.5. Thalamocortical and the earliest corticofugal (preplate) projections subsequently meet and begin to fasciculate with each other in the primitive internal capsule, under the anlage of the striatum. At E15.5 thalamocortical projections reach the appropriate cortical areas by associating to the early corticofugal fibre scaffold. CTX, cerebral cortex; DT, dorsal thalamus; PIC, primitive internal capsule. (B) In the L1 knockout mice the initial steps are indistinguishable. Subsequently fasciculation abnormalities occur and some thalamic projections get misrouted at the striatocortical junction. The fascicles formed by thalamocortical axons in the striatum appear thicker and more disorganized. Some fibre fascicles get derailed at the striatocortical junction and defasciculate or turn abruptly within the striatum, and do not seem to reach the cortex. (C) In *reeler* mouse, at E13, the formation of the preplate, the outgrowth of descending corticofugal axons and the concomitant growth of thalamic axons through the primitive internal capsule all occur indistinguishably in *reeler* and normal animals. But by E14.5, as thalamic axons are approaching their target areas, the cortical plate itself has started to form and the distinctive differences in the phenotypes begin to emerge. In *reeler* the cortical plate forms under all the preplate (now superplate) cells, whose pioneer axons gather into oblique fascicles, running through the thickening cortical plate. Thalamic axons follow these fibres up to the superplate layer. In *reeler* at E14.5, thalamic fibres grow up obliquely through the cortical plate and some have already entered the superplate. At E15.5, in *reeler* thalamic axons all appear to have passed through the plate and into the superplate above. This behaviour can only be observed in normals in a few occasions, whereas almost all axons show this behaviour in *reeler*. Modified from Molnár & Hannan (2000).

which can bind L1 *in vitro* (Friedlander et al 1994). Fukuda et al (1997) showed that subplate projections express neurocan immunoreactivity, that L1 immunoreactivity is specifically localized on the growing thalamic axon in embryonic rat brain, and that L1-bearing thalamocortical axons extend along neurocan-positive fibres.

Recently, a transgenic mouse that lacks L1 expression has been generated (Cohen et al 1997). Carbocyanine dye tracing experiments at embryonic and early postnatal stages reveal altered fasciculation at the striatocortical boundary (Molnár et al 1999). In the mutant, thalamic projections travel abnormally through the striatum in large bundles. Some fibres form aberrant projections into the striatum and do not make it to the cortex (Fig. 2B). Nevertheless, the majority of the thalamic fibres branch and arborize normally in the cortical plate and assume a periphery-related pattern in the primary somatosensory cortex. In the L1 knockout mouse the normal whisker-related pattern is confirmed with cytochrome oxidase and 2-deoxyglucose experiments, and the presence of barrels

---

FIG. 2.   Schematic summary diagram of the early connectivity revealed with carbocyanine dye tracing in embryonic forebrain at embryonic day (E) 14.5. Each large dot represent a crystal placement site in the dorsal (A) and ventral (B) thalamus, in medial (C) and lateral (D) part of the internal capsule, in dorsal (E) or perirhinal cortex. Small dots indicate the distribution of back-labelled cells. Arrows indicate regions where abrupt changes were observed in connections, indicating possible subdivisions and boundaries. A and B demonstrate the location of cells which have projections to the dorsal and ventral thalamus: cells in the epithalamus, thalamic reticular nucleus, hypothalamus and in lateral and medial internal capsule (below the lateral and medial ganglionic eminences). The first fibres reaching the dorsal and ventral thalamus originate from cells in the internal capsule. C and D give a summary of the distribution of the cells labelled from crystal placements in the medial and lateral parts of the internal capsule. Labelled cells were observed in the internal capsule and perirhinal cortex in dorsal but not ventral thalamus. At E15 preplate (marginal zone and subplate) cells were labelled as the front of labelled cells extended into dorsal cortex. These results indicate that projections from the preplate cells reach the internal capsule by E14.5–15 and run in the region of the perireticular cells, where they pause between E14–15. The corticofugal projections do not reach dorsal or ventral thalamus until sometime later at E16. At E15, after the first true cortical plate cells have migrated into the lateral cortex, some cells of the cortical plate, as well as preplate cells (now forming the subplate and marginal zone), are back-labelled from the lateral or medial parts of the internal capsule, but still not from the dorsal thalamus, even with long incubation periods. E and F demonstrate the cells labelled from the dorsal cerebral cortex (E) and from the perirhinal cortex (F). At E14 and E15 crystal placements into the dorsal cortex revealed extensive long-range connections within the preplate. At E14 and E15 perirhinal cortical crystal placements labelled internal capsule cells, whereas dorsal cortical crystal placements did not. From dorsal cortex, few or no internal capsule cells were labelled, whereas from perirhinal cortex a small number of cells were consistently labelled. (Modified from Molnár & Cordery 1999).

is demonstrated with Nissl staining (Molnár et al 1999). It is yet to be examined whether the topographical order is maintained in spite of the altered trajectories or reinstalled along the path of thalamocortical projections in the mutant.

Analysis of thalamocortical development in *reeler* mutant mouse proved to be extremely useful in testing ideas on selective fasciculation of thalamic axons (Molnár et al 1998c). In the mutant, despite abnormalities in the position of cortical layers, thalamic fibres reach the correct regions of cortex and terminate on neurons equivalent to layer IV (Caviness et al 1988). The existence of highly permissive, privileged pathways for axon growth along early corticofugal projections could explain how thalamic axons in *reeler* are able to penetrate the cortical plate in fascicles and steer up to reach the equivalent cells in the

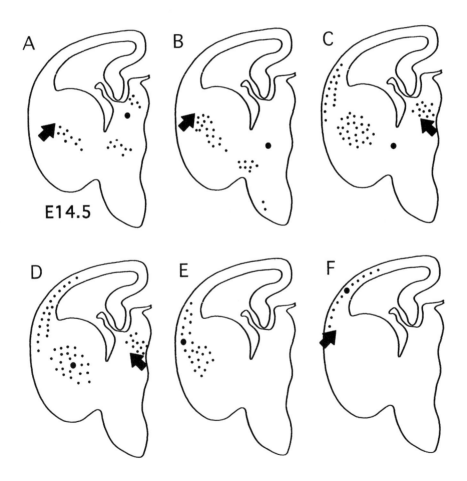

superplate, while ignoring the hostile territory of cortical plate cells around them (Fig. 1C). The peculiar local pattern of innervation seen in the adult is compatible with, and provides good evidence for, the theory of axon guidance developed in normal rodents (Molnár & Blakemore 1995, Molnár 1998).

Recently, Hevner and colleagues (1998) showed errors of corticothalamic and thalamocortical pathfinding in various transgenic mice with mutations of transcription factor genes. The abnormalities occurred in the region of the internal capsule, away from regions where the affected genes are normally expressed, in either cortex (*Tbr1*), dorsal thalamus (*Gbx2*), or both (*Pax6*). In all the three mutants the failures of the thalamocortical pathfinding were always accompanied with abnormalities of the corticofugal projections; both projections failed to cross the internal capsule. Hevner et al (1998) proposed that these results indicate that the interaction of thalamocortical projections with their early corticofugal counterparts is necessary for the normal development of both sets of projections.

Possible early boundaries restricting regional connectivity in the mammalian forebrain

Thalamic projections begin to grow towards the cortex at an early stage when the majority of cells destined for the subdivisions of forebrain have yet to be generated. The early generated and largely transient cells in the thalamic reticular nucleus, internal capsule and preplate initially form a continuum in the developing forebrain along the future path of thalamic and cortical projections. Nevertheless, there are clear differences emerging in their connections, suggesting some form of early partitioning in the forebrain. Connectional analysis with carbocyanine dye tracing revealed sudden changes in the connectivity, indicating anatomical subdivisions with possible boundaries (i) between the cortical intermediate zone and the ganglionic eminence, (ii) between the dorsal and the ventral thalamus and (iii) between perirhinal and dorsal cortex (Fig. 2; Molnár & Cordery 1999):

(1)   Tracing from diencephalon revealed a continuous chain of cells extending from the dorsal thalamus to the ganglionic eminence, but none in the cerebral wall. The front of back-labelled cells respects a border slightly medial to the corticostriatal junction, extending between the striatum and the intermediate zone (Fig. 2A, B).

(2)   Tracing from the medial or lateral internal capsule revealed back-labelled cells in the ganglionic eminence and in dorsal thalamus, but never in ventral thalamus (Fig. 2C, D). Although many of the ventral thalamic cells project to dorsal thalamus, they avoid developing projections out of the diencephalon into the internal capsule or the ganglionic eminence. At embryonic day (E) 14 DiI (1,1′dioctadecyl-3,3,3′,3′-tetramethylindocarbocyanine perchlorate)

crystal placements in the internal capsule labelled cells in perirhinal cortex, but few in the cerebral cortex.

(3) Crystal placement into the dorsal cortex labelled no internal capsule cells (until E17, see Fig. 2E), in contrast to the perirhinal cortex, where a small number of cells were consistently revealed from a similar crystal placement from E14–16 (Fig. 2F).

Puzzling features at the striatocortical junction in early mammalian pallium

The above subdivisions might provide the first means of functional partitioning of the pallium, and developing connections might require a sequence of precise interactions with early generated cells and their projections to be able to advance towards their correct targets (see Mitrofanis & Guillery 1993, Molnár & Blakemore 1995, Métin & Godement 1996). During early stages of forebrain development, thalamocortical and corticofugal projections pause before they cross the striatocortical junction (Métin & Godement 1996, Clascá et al 1995, Molnár et al 1998a, Molnár & Cordery 1999), and their subsequent interaction was proposed to occur exactly at this region of the primitive internal capsule, beyond which thalamic fibres cofasciculate with early corticofugal projections to reach cortex (Molnár & Blakemore 1990, 1995, Molnár et al 1998a). Thalamocortical projections show a distinct change in their fasciculation pattern at this border. The fascicles are reassembled into individual fibres as they enter the intermediate zone (Molnár et al 1998b; Fig. 3C). This is slightly lateral from the group of cells in the internal capsule which develop early projections to the dorsal thalamus.

Corresponding to this zone at the striatocortical junction lies a stripe of cells that express various genes distinct from the intermediate zone or from the anlage of the striatum. Liu & Graybiel (1992) and Métin & Godement (1996) described a narrow and dense stripe of calbindin- and MAP2-positive neurons at the lateral border of the lateral ganglionic eminence in rat and hamster embryos. Other markers reveal a similar stripe of cells extending from the ventricular zone of the striatocortical junction to the ventral telencephalon (Table 1). The staining patterns coincide with the region where thalamocortical and corticothalamic projections pause during their development and where thalamocortical projections undergo a drastic change in their fasciculation pattern. Hevner et al (1998) and Kawano et al (1999) showed that both sets of fibres fail to traverse this region in the *Tbr1*, *Gbx2* and *Pax6* knockout mice. It will be important to understand more of the cellular and molecular interactions in this area so as to be able to explain why this region is critical for thalamocortical development. Perhaps the stripe of cells is thicker and impenetrable to early corticofugal and thalamocortical fibres due to delayed or disturbed migration of cells, or due to changed expression of surface molecules in these mutants. What is the origin and fate of this transient stripe of cells at the gateway of the internal capsule?

Comparative developmental studies help to reveal homologous structures and developmental patterns

The understanding of the mechanisms of each step during the development, though itself a considerable challenge, is far simpler than the explanation of why Nature uses a particular solution for a particular developmental step. The comparative analysis of development is one approach to these questions. Comparison of immature stages across species reveals features that are otherwise obstructed by the complexity of the mature brain. Comparisons could be made easier by searching for homologies in developing brains. While looking at development in evolutionary terms helps to focus on the most biologically relevant mechanisms (Raff 1996, Butler & Molnár 1998), conversely

FIG. 3.   The striatocortical junction in embryonic day (E) 14–17 rat brains. The sections in A, B and C were counterstained with bisbenzimide and examined in a fluorescent microscope under ultraviolet illumination (appear blue), and with filters to reveal the DiI (1,1′dioctadecyl-3,3,3′,3′-tetramethylindocarbocyanine perchlorate) (orange-red) and DiAsp (yellow-green) labelling. The dark field photomicrographs of A, B and C were taken with multiple exposures to superimpose two (B, C) or three (A) fluorescent images obtained with different fluorescent filters. (A) Within the same hemisphere of an E15 brain a single DiI crystal was placed into the dorsal cortex (to reveal the descending corticofugal projections) and a single DiAsp crystal was placed into the dorsal thalamus (to reveal the internal capsule cells). After four weeks of incubation at room temperature, sections were cut 45° to the coronal plane to be able to follow the fibres for long distances within a single section. Early corticofugal projections extend towards the region of the perireticular cells, where they do not enter deep into the ganglionic eminence as they arrive to the lateral entrance of the internal capsule. They seem to pause at this location between E14–15. The blackballed cells in the internal capsule (appear yellow) are situated within and around the internal capsule and their lateral boundary seems to correlate to the front of the corticofugal fibre tips. (B) A small crystal of DiI was implanted into the ventral thalamus of a fixed E14 brain and after four weeks incubation at 37 °C, 100 μm-thick coronal sections were cut and double exposure photomicrographs were taken using appropriate filters for DiI and bizbenzimide. The tracing revealed numerous back-labelled cells in the primitive internal capsule below the ganglionic eminence. (C) Radial glia labelled from a DiI crystal placement to the olfactory bulb at E17. The radial glia end feet are located slightly medial to the striatocortical junction (upper right corner). The glia processes originate from here and descend ventrolaterally along the curvature of the striatocortical junction to reach the olfactory cortex (towards lower left corner, not shown). (D) Calretinin immunohistochemistry in a E16 rat brain revealed cells in the marginal zone, subplate and in the primitive internal capsule. The calretinin-positive cells of the internal capsule have similar location than the back-labelled cells presented in A and B. CP, cortical plate; CTX, cerebral cortex; GE, ganglionic eminence; MZ, marginal zone; PIC, primitive internal capsule; SP, subplate. Arrows at A, B and D point to the lateral front of the internal capsule cells in corresponding position. Bars (A, B and D) = 200 μm. Bar (C) = 100 μm.

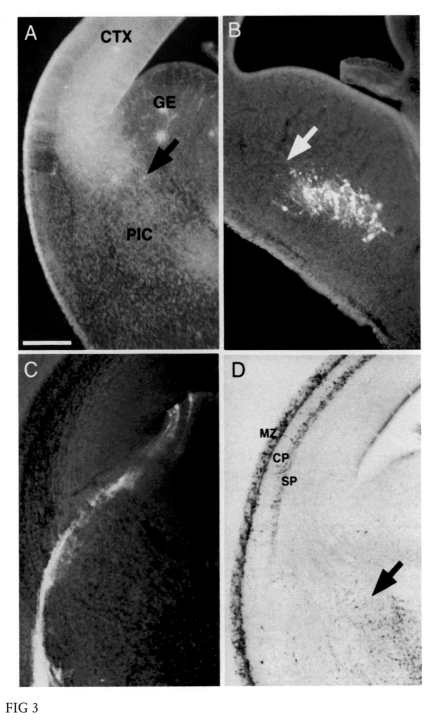

FIG 3

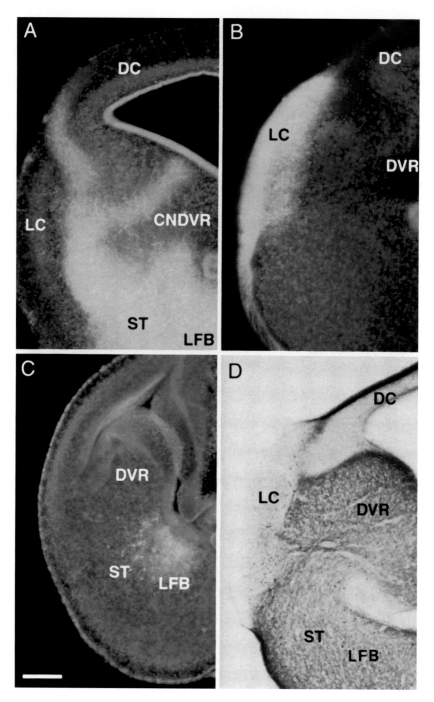

FIG 4

developmental studies can reveal evolutionary links important because some of the participants of the early interactions are transient; a large proportion of them disappear by adulthood. However, it is thus far unknown whether non-mammalian species, particularly reptiles, possesses the equivalent groups of subplate and Cajal-Retzius cells and whether they subserve similar or different functions in these species.

Development of thalamocortical projections in reptiles

Chelonians are generally believed to be derived from ancestors closest to the stem reptiles, which also gave rise to synapsids and eosuchians. Therefore, they could provide an interesting system for studying conserved developmental mechanisms because they might show more similarities to the presumed ancestors from which the major phylogenetic categories started to diverge. We set out to compare the development of thalamocortical projections in embryonic turtle with mammals to gain insight into the early mechanisms of forebrain development (Molnár & Cordery 1999, Cordery & Molnár 1999). In turtle embryos, fibres from dorsal thalamus extend through the lateral forebrain bundle and striatum to reach their targets in dorsal cortex and dorsal ventricular ridge (DVR) from stage 19–20 (Fig. 4A). The labelled fibres extend along specific paths among back-labelled cells in the ventral thalamus and lateral forebrain bundle to reach their major targets: the ventral and dorsal part of the DVR and the molecular layer of the dorsal cortex (Fig. 4A). Thalamic fibres avoid regions of the central area of DVR, lateral cortex or the lower cellular and subcellular layers of dorsal cortex. A distinct

---

FIG. 4. Dorsal and lateral cortex, dorsal ventricular ridge (DVR) and striatum in embryonic turtle at stage 20–22. All sections were cut in coronal plane. The fluorescent micrographs of A, B and C were taken with double exposure using filters for DiI (1,1'dioctadecyl-3,3,3',3'-tetramethylindocarbocyanine perchlorate) (red-orange) and bisbenzimide (blue) counter-staining. (A) Thalamocortical fibres revealed with DiI crystal placement to the dorsal thalamus in a stage 25 turtle embryo. The counterstaining shows the cell patterning in dorsal cortex and the DVR. Numerous DiI-labelled fibres enter the DVR in two major zones, but the core nucleus is relatively free of fibre label. Some thalamic fibres extend to the surface of the dorsal cortex where they remain restricted to the marginal zone. (B) Olfactory projections in the lateral cortex of a stage 22 turtle. The olfactory projections were labelled from a crystal placement to the olfactory bulb. The lateral cortex is densely innervated. The dorsal cortex and the core nucleus of the DVR do not receive olfactory projections. (C) Ventral thalamic DiI placement in a stage 20 turtle brain numerous labelled cells in the striatum and lateral forebrain bundle indicating that their projection reached the thalamic crystal placement site. (D) Calretinin staining in an embryonic stage 20 turtle telencephalon. Calretinin positive cells can be seen in the lateral forebrain bundle, striatum, DVR and lateral cortex, but not within the dorsal cortex itself. The dorsal and lateral cortex both have relatively pale neuropil staining. CNDVR, core nucleus of dorsal ventricular ridge; DC, dorsal cortex; LC, lateral cortex; LFB, lateral forebrain bundle; ST, striatum. Bars (A, B and D) = 100 μm, Bar (C) = 200 μm.

pattern of innervation is apparent from stage 20, from the earliest stage at which thalamic fibres could be back-labelled. There is clear segregation of thalamic fibres originating from nucleus rotundus and the perirotundal complex to DVR and to dorsal cortex from embryonic stages (Cordery & Molnár 1999) as described in adults (Hall & Ebner 1970). The specific developmental process most probably involves intrinsic guidance mechanisms, possibly including guide-post cells in the lateral forebrain bundle and striatum, similar to transient internal capsule cell populations in mammals.

We were particularly interested in two distinct cell groups which might have a specific role in the deployment of thalamic fibres into and through the internal capsule. The first group is the group of internal capsule cells which develop early thalamic projections, and the second is a stripe of cells extending from the ventricular zone of the striatocortical junction to ventral telencephalon. The presence of the transient cells in internal capsule was described in several mammalian species including rodents, carnivores, marsupials and humans (Mitrofanis 1992, Molnár et al 1998a,b, Letinic & Kostovic 1996). Their early thalamic projections were described by Métin & Godement (1996) in hamster and by Molnár et al (1998a,b) in rodents and marsupials. The data in turtle embryos are consistent with the possibility that the cells in the lateral forebrain bundle and striatum are homologous to the ones observed with thalamic projections in the embryonic internal capsule in hamster (Métin & Godement 1996), rat (Molnár et al 1998a, Molnár & Cordery 1999), and the marsupial *Monodelphis domestica* (Molnár et al 1998c). This suggestion is based on connectional analysis (Molnár & Cordery 1999, Cordery & Molnár 1999), and is supported by the similarity of the morphology of the back-labelled cells and matching immunoreactivity for calretinin and NPY in the corresponding regions of embryonic rat and embryonic turtle brains (Cordery et al 1997). The transient stripe of cells was proposed to lie at the striatocortical junction lateral to these cells in the internal capsule, and they have no projections to the dorsal thalamus. There are several markers whose expression mark their position at embryonic stages (Table 1).

Comparison of embryonic pallial organization in reptiles and mammals

Based on our hodological analysis in rat embryos we suggested that there is a sudden change in the properties of the projection neurons between the lateral edge of the striatum and the cortical intermediate zone (Molnár & Cordery 1999 and Fig. 2). In turtle embryos, there appears to be a similar change (defined by our connectional analysis, Cordery & Molnár 1999) as no cells were labelled from dorsal thalamic crystal placement beyond the striatum and ventral

TABLE 1   Gene expression of the stripe of cells at the striatocortical junction (A) and in cells of the primitive internal capsule (B)

A

Gene	Reference
Calbindin	Liu & Graybiel (1992)
MAP2	Métin & Godement (1996)
CAD11	Simonneau & Thiery (1998)
Semaphorin G	Skaliora et al (1998)
Pax6 positive stripe	Stoykova et al (1997) Fernández et al (1998)
Emx1-negative, *Dlx*-negative stripe (intermediate zone)	Fernández et al (1998)
β-galactosidase expression driven by tubulin promoter	R. Adams & J. Nangla (unpublished observation 1998)
Staining with Pruss reaction[a]	C. Métin & N. Ropert (unpublished observations (1998)

[a]Based on Ca^{2+} permeability of AMPA receptors.

B

Gene	Reference
Pro-a-thyrotropin releasing hormone	Mitrofanis (1992)
γ-aminobutyric acid (GABA), somatostatin, parvalbumin	Mitrofanis (1992)
Calbindin, calretinin	Mitrofanis (1992), Amadeo et al (1998)
CAT-301	Crabtree & Kind (1993)
Semaphorin D	Skaliora et al (1998)
Dlx1	Fernández et al (1998)
NETRIN-1	Métin et al (1997)

part of the anterior dorsal ventricular ridge (ADVR) (Fig. 4C). It could correspond to a matching boundary extending between the striatum and ventral part of DVR.

This suggestion is supported by early gene expression patterns of Fernández and colleagues (1998). They examined the developmental expression pattern of homeobox genes of the *Emx*, *Dlx* and *Pax* families in embryonic forebrains of mouse, chick, turtle and frog and described that at early stages of neurogenesis the expression domains of *Emx1* in the pallium and *Dlx1* in the striatum are separated by a thin intermediate territory expressing neither gene in all examined species. This intermediate stripe expressed *Pax6*. In turtle this intermediate domain contained the ridge of the ADVR, but was not coextensive with it. Fernández et al (1998) proposed that the homologue of DVR of reptiles and the neostriatum of birds might be related to the lateralobasal part of the amygdala which also corresponds to the intermediate compartment. We were interested in the possibility that the transient cells within the mammalian ganglionic eminence are somehow related to the DVR in reptiles. Our tracing study from the dorsal thalamus in mammals suggests that most of the back-labelled internal capsule cells are located medial to the region of the intermediate territory defined by Fernández et al (1998) and do not extend into it. Similarly in reptiles ADVR (which has similar developmental gene expression pattern to the mammalian intermediate zone; Fernández et al 1998), does not contain back-labelled cells after thalamic crystal placements. This suggests that although the cortical intermediate zone in mammals and the ventral border of DVR in reptiles are close to the front of the back-labelled cells in the internal capsule, they are excluded from these structures (Cordery & Molnár 1999). It is likely that the stripe of cells extending to the striatocortical junction is related to the anlage of DVR in reptiles as suggested by Fernández et al (1998), and the group of cells in the internal capsule is homologous to some cells of the lateral forebrain bundle and of striatum in turtle (Cordery & Molnár 1999). The developmental role of the two groups of cells is not yet established. The internal capsule cells with early thalamic projections might play a role in the early outgrowth of thalamic fibres or in the sorting of corticofugal projections to thalamus or cerebral peduncle (Mitrofanis & Guillery 1992). They could also provide temporary targets (Métin & Godement 1996). The transient stripe of cells (Figs 2 and 4) extending along the striatocortical junction might be responsible for the behaviour of thalamocortical and especially corticothalamic fibres. These cells might transiently obstruct the gateway between thalamus and cortex during early embryonic development in mammals. However, whether this stripe is indeed an obstacle for thalamocortical and corticofugal axons has yet to be established.

Open questions on the developmental role and evolutionary origin of the transient cells in the mammalian primitive internal capsule

Perhaps the developmental steps observed in mammals might be common in numerous vertebrates and represent an evolutionary conserved blueprint. It is of interest to define what is the homologue of the two groups of cells in the internal capsule in other vertebrates. The homology of the lateral cortical pallial region in all amniotes is relatively well established and non-controversial (see Butler & Hodos 1996). In contrast, the evolutionary relationship of the DVR, a large pallial region in diapsid reptiles, birds, and turtles to some part of the mammalian telencephalon remains in dispute. Homology of the ADVR with various parts of mammalian neocortex was originally proposed by Karten (1969) based on studies of auditory and visual pathways to the DVR from the midbrain roof via the respective dorsal thalamic nuclei. Recently, several non-neocortical telencephalic structures have been proposed as homologues of the DVR. Bruce & Neary (1995) argued that the anterior and posterior parts of the DVR are homologous to the lateral and basolateral amygdala of mammals, respectively. Striedter (1997) argued that while the posterior DVR may be homologous to both the lateral and basolateral (i.e. pallial) amygdala, the ADVR is homologous to the mammalian endopiriform nucleus, and the pallial thickening (as present in turtles) is homologous to the mammalian claustrum proper, as Holmgren (1925) had previously suggested. Early gene expression and cell lineage studies will eventually resolve the issue of whether the cells contributing to amygdala are generated at a similar site of the embryonic neuroepithelium as the cells of the reptilian DVR. It is interesting to note that in embryonic turtle forebrain, thalamic fibres avoid the core of DVR during early development, and parts of DVR are considered to be homologous to the intermediate zone corresponding to the transient stripe of cells (Fernández et al 1998). If most of the embryonic markers are expressed similarly in mammals and turtles, perhaps corresponding surface molecules have conserved expression in homologous structures in mammals and reptiles. Perhaps the stripe of cells extending along the striatocortical junction in mammals is not permissive for thalamocortical and corticofugal fibre growth at early stages of development. In embryonic turtles, the core nucleus of the DVR, which was proposed to be the reptilian homologue of the cells constituting the transient stripe, has much less dense innervation compared to the surrounding regions at early stages of development. An interesting question that remains is how would mammalian thalamic fibres respond to reptilian DVR and dorsal cortical tissue?

Conclusions and hypotheses on the conserved mechanisms

In all examined mammals, early development follows a similar general pattern. Thalamocortical projections have to pass through several emerging telencephalic subdivisions to reach the cortex. Descriptive studies of thalamocortical development suggested that there are marked changes in the properties of the extracellular environment which are reflected in the kinetics of growth and fasciculation patterns of thalamocortical projections. A puzzling behaviour of thalamocortical and corticothalamic projections is observed at the internal capsule at the boundary slightly more medial than the junction between cortical intermediate zone and the ganglionic eminence. This region appears to be critical for thalamocortical development. During early stages of forebrain development, thalamocortical and corticofugal projections pause before they cross this boundary, and their subsequent interaction was proposed to occur at this region of the primitive internal capsule, beyond which thalamic fibres cofasciculate with early corticofugal projections to reach the cortex. Studies on various mutant and transgenic mice demonstrate altered or disrupted thalamocortical development at this particular site.

Hodological analysis in mammalian embryonic brains reveals that thalamic reticular cells and some cells of the primitive internal capsule project to the dorsal thalamus from early embryonic ages. These internal capsule cells are distinct from a transient stripe of cells at the striatocortical junction, revealed by their distinct gene expression pattern and lack of thalamic projections. It is proposed that the possible function during development and evolutionary origin of the two groups of cells are different. The internal capsule cells with thalamic projections might be responsible for the early outgrowth of thalamic fibres and perhaps for the sorting of various corticofugal projections. Their likely homologue is a group of cells in the reptilian lateral forebrain bundle and striatum. One possibility is that the transient stripe of cells temporarily blocks the gateway between the cortex and thalamus, causing the pause and the subsequent fasciculation of subplate projections with thalamic projections. This group of cells in mammals might be the homologue of the DVR in reptiles. There are several unresolved issues in thalamocortical development in both developmental and evolutionary biology, nevertheless we are beginning to understand the general patterns of cellular and

---

FIG. 5.   A schematic diagram of the thalamocortical and olfactory projections (upper panels) and the location of possible homologous cell groups in embryonic turtle (stage 20) and rat (embryonic day [E] 14.5) pallium (lower panels). Thalamocortical projections pause at the lateral part of the primitive internal capsule (PIC) before eventually passing through the region in large fascicles. The projections continue their growth as individual fibres in the cortical intermediate zone (IZ). In turtles, thalamic projections extend along specific paths to reach the dorsal and ventral part of the dorsal ventricular ridge (DVR) and superficial layer of the dorsal cortex, but avoid the core of DVR (CNDVR) and the ventral layers of dorsal and the entire olfactory cortex. Connectional analysis in embryonic rat and turtle and gene expression patterns suggest that a similar boundary (marked in lower panels) exist between striatum and the cortical IZ (in mammals) and the ventral part of the DVR and the striatum (in turtle). A boundary exists in the thalamus in mammals between thalamic reticular nucleus (TRN) and dorsal thalamus and this border also seem to exist in the turtle between dorsal and ventral thalamus. It is yet to be determined whether the internal capsule (IC) and lateral forebrain bundle (LFB) cells can be subdivided into numerous subgroups and what is the homologous cell group of the reptilian DVR in mammals. DC, dorsal cortex; DT dorsal thalamus; LC, lateral cortex; MZ, marginal zone; PRC, perithinal cortex; SP, subplate; ST, striatum; VT, ventral thalamus.

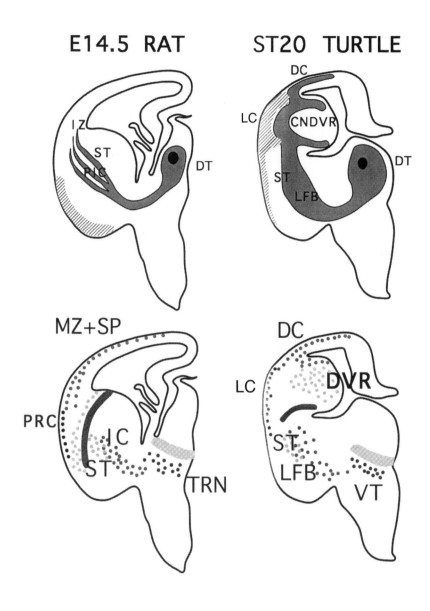

molecular interactions and the underlying logic of common developmental algorithms.

Acknowledgements

I would like to thank Patricia Cordery, Richard Adams and Colin Blakemore whose contributions I discussed here. I am indebted to Graham Knott, David Price, Egbert Welker, Dan Lavery and Hendry Kennedy for discussions and for their thoughtful comments of an earlier version of this manuscript. Thanks goes to Erik Bernardi and Mark Berney for their expert help with photography. The original work of the author was supported by Medical Research Council, UK (G9706008) and Swiss National Science Foundation (3100-51036.97 and 3100-56032.98).

References

Allendoerfer KL, Shatz CJ 1994 The subplate, a transient neocortical structure: its role in the development of connections between thalamus and cortex. Annu Rev Neurosci 17:185–218

Amadeo A, De Biasi S, Frassoni C, Ortino B, Spreafico R 1998 Immunocytochemical and ultrastructural study of the rat perireticular thalamic nucleus during postnatal development. J Comp Neurol 392:390–401

Bicknese AR, Sheppard AM, O'Leary DDM, Pearlman AL 1994 Thalamocortical axons extend along a chondroitin sulfate proteoglycan-enriched pathway coincident with the neocortical subplate and distinct from the efferent path. J Neurosci 14:3500–3510

Blakemore C, Molnár Z 1990 Factors involved in the establishment of specific interconnections between thalamus and cerebral cortex. Cold Spring Harbor Symp Quant Biol 55:491–504

Bruce LL, Neary TJ 1995 The limbic system of tetrapods: a comparative analysis of cortical and amygdalar populations. Brain Behav Evol 46:224–234

Butler AB, Hodos W 1996 Comparative vertebrate neuroanatomy: evolution and adaptation. Wiley-Liss, New York

Butler AB, Molnár Z 1998 Development and evolution in nervous systems: development and evolution of ideas. Trends Neurosci 21:177–178

Catalano SM, Robertson RT, Killackey HP 1991 Early ingrowth of thalamocortical afferents to the neocortex of the prenatal rat. Proc Natl Acad Sci USA 88:2999–3003

Catalano SM, Robertson RT, Killackey HP 1996 Individual axon morphology and thalamocortical topography in developing rat somatosensory cortex. J Comp Neurol 367:36–53

Caviness VS Jr, Crandall JE, Edwards MA 1988 The *reeler* malformation, implications for neocortical histogenesis. In: Jones EG, Peters A (eds) Cerebral cortex, vol 7: Development and maturation of cerebral cortex. Plenum Press, New York, p 59–89

Chung WW, Lagenaur CF, Yan YM, Lund JS 1991 Developmental expression of neuronal cell adhesion molecules in the mouse neocortex and olfactory bulb. J Comp Neurol 314:290–305

Clascá F, Angelucci A, Sur M 1995 Layer-specific programs of development in neocortical projection neurons. Proc Natl Acad Sci USA 92:11145–11149

Cohen NR, Taylor JSH, Scott LB, Guillery RW, Soriano P, Furley AJW 1997 Errors in corticospinal axon guidance in mice lacking the neural cell adhesion molecule L1. Curr Biol 8:26–33

Cordery P, Molnár Z 1999 Embryonic development of connections in turtle pallium. J Comp Neurol 413:26–54

Cordery P, Blakemore C, Molnár Z 1997 Comparison of mammalian and reptilian pallium during development. Soc Neurosci Abstr 23:320.19

Crabtree JW, Kind PC 1993 Monoclonal antibody Cat-301 selectively identifies a subset of nuclei in the cat's somatosensory thalamus. J Neurocytol 22:903–912

De Carlos JA, O'Leary DDM 1992 Growth and targeting of subplate axons and establishment of major cortical pathways. J Neurosci 12:1194–1211

Derer P, Nakanishi S 1983 Extracellular matrix distribution during neocortical wall ontogenesis in normal and *reeler* mice. J Hirnforsch 24:209–224

Erzurumlu RS, Jhaveri S 1992 Emergence of connectivity in the embryonic rat parietal cortex. Cereb Cortex 2:336–352

Fernández AS, Pieau C, Repérant J, Boncinelli E, Wassef M 1998 Expression of the *Emx-1* and *Dlx-1* homeobox genes define three molecularly distinct domains in the telencephalon of mouse, chick, turtle and frog embryos: implications for the evolution of telencephalic subdivisions in amniotes. Development 101:2099–2111

Friedlander DR, Milev P, Karthikeyan L, Margolis RK, Margolis RU 1994 The neuronal chondroitin sulfate proteoglycan neurocan binds to the neural cell adhesion molecules Ng-CAM/L1/NILE and N-CAM, and inhibits neuronal adhesion and neurite outgrowth. J Cell Biol 125:669–680

Fukuda T, Kawano H, Ohyama K et al 1997 Immunohistochemical localization of neurocan and L1 in the formation of thalamocortical pathway of developing rats. J Comp Neurol 382:141–152

Godfraind C, Schachner M, Goffinet AM 1988 Immunohistological localisation of cell adhesion molecules L1, J1, N-CAM and their common carbohydrate L2 in the embryonic cortex of normal and *reeler* mice. Brain Res 470:99–111

Hall WC, Ebner FF 1970 Thalamotelencephalic projections in the turtle (*Pseudemys scripta*). J Comp Neurol 140:101–122

Hevner RF, Miyashita E, Martin G, Rubenstein JLR 1998 Lack of thalamocortical connections in mutants affecting cortical (TBR-1) or thalamic (GBX-2) gene expression. Soc Neurosci Abstr 24:58

Holmgren N 1925 Points of view concerning forebrain morphology in higher vertebrates. Acta Zool Stockh 6:413–477

Karten HJ 1969 The organization of the avian telencephalon and some speculations on the phylogeny of the amniote telecephalon. Ann NY Acad Sci 167:164–179

Kawano H, Fukuda T, Kubo K et al 1999 Pax-6 is required for thalamocortical pathway formation in fetal rats. J Comp Neurol 408:147–160

Letinic K, Kostovic I 1996 Transient neuronal population of the internal capsule in the developing human cerebrum. Neuroreport 7:2159–2162

Liu F-C, Graybiel AM 1992 Transient calbindin-D28K-positive systems in the telencephalon: ganglionic eminence, developing striatum and cerebral cortex. J Neurosci 12:674–690

McConnell SK, Ghosh A, Shatz CJ 1989 Subplate neurons pioneer the first axon pathway from the cerebral cortex. Science 245:978–982

Métin C, Godement P 1996 The ganglionic eminence may be an intermediate target for corticofugal and thalamocortical axons. J Neurosci 16:3219–3235

Métin C, Deléglise D, Serafini T, Kennedy TE, Tessier-Lavigne M 1997 A role for netrin-1 in the guidance of cortical efferents. Development 124:5063–5074

Miller B, Sheppard AM, Bicknese AR, Pearlman AM 1995 Chondroitin sulfate proteoglycans in the developing cerebral cortex: the distribution of neurocan distinguishes forming afferent and efferent axonal pathways. J Comp Neurol 355:615–628

Mitrofanis J 1992 Patterns of antigenic expression in the thalamic reticular nucleus of developing rats. J Comp Neurol 320:161–181

Mitrofanis J, Guillery RW 1993 New views of the thalamic reticular nucleus in the adult and developing brain. Trends Neurosci 16:240–245

Molnár Z 1998 Development of thalamocortical connections. Springer-Verlag, Berlin

Molnár Z, Blakemore C 1990 Relationship of corticofugal and corticopetal projections in the prenatal establishment of projections from thalamic nuclei to the specific cortical areas of the rat. J Physiol 430:104

Molnár Z, Blakemore C 1995 How do thalamic axons find their way to the cortex? Trends Neurosci 18:389–397

Molnár Z, Cordery PM 1999 Connections between cells of the internal capsule, thalamus and cerebral cortex in embryonic rat. J Comp Neurol 413:1–25

Molnár Z, Hannan A J 2000 Development of thalamocortical projections in normal and mutant mice. In: Goffinet AM, Rakic P (eds) Development of the cerebral cortex of the mouse. Springer-Verlag, Heidelberg, in press

Molnár Z, Adams R, Blakemore C 1998a Mechanisms underlying the establishment of topographically ordered early thalamocortical connections in the rat. J Neurosci 18:5723–5745

Molnár Z, Adams R, Goffinet AM, Blakemore C 1998b The role of the first postmitotic cells in the development of thalamocortical fibre ordering in the *reeler* mouse. J Neurosci 18:5746–5785

Molnár Z, Knott GW, Blakemore C, Saunders NR 1998c Development of thalamocortical projections in the South American gray short-tailed opossum (*Monodelphis domestica*). J Comp Neurol 398:491–514

Molnár Z, Mather NK, Katsnelson A et al 1999 Disturbed fasciculation, but ordered cortical termination of thalamocortical projections in L1 knockout mice. Soc Neurosci Abstr 25:526

Raff R A 1996 The shape of life: genes, development and the evolution of animal form. University of Chicago Press, Chicago, IL

Simonneau L, Thiery JP 1998 The mesenchymal cadherin-11 is expressed in restricted sites during the ontogeny of the rat brain in modes suggesting novel functions. Cell Adhesion Commun 6:431–450

Skaliora I, Singer W, Betz H, Püschel AW 1998 Differential patterns of semaphorin expression in the developing rat brain. Eur J Neurosci 10:1215–1229

Stewart GR, Pearlman AL 1987 Fibronectin-like immunoreactivity in the developing cerebral cortex. J Neurosci 7:3325–3333

Stoykova A, Götz M, Gruss P, Price J 1997 Pax-6-dependent regulation of adhesive patterning, R-cadherin expression and boundary formation in developing forebrain. Development 124:3765–3777

Striedter GF 1997 The telencephalon of tetrapods in evolution. Brain Behav Evol 49:179–213

DISCUSSION

Rubenstein: What is the nature of the early cells in the striatum that project to the reticular nucleus and the dorsal thalamus?

Molnár: The reason why I call them internal capsule cells is because the identity of these cells is not known. It is possible that we shall be able to divide them into subgroups with future work. John Mitrofanis and Ray Guillery (1993) described these cells, and they have given the name, perireticular nucleus, because they believed that they are diencephalic in origin and are different from striatal neurons. I find it difficult to distinguish a real nucleus at early stages without additional markers. Mitrofanis (1992) identified the perireticular neurons based on their smaller size and distinct immunohistochemistry. Unfortunately these criteria cannot always be applied in combination with DiI (1,1'dioctadecyl-3,3,3',3'-tetramethylindocarbocyanine perchlorate) labelling from early embryonic age and until it is not established that the same internal capsule cells have similar characteristics as the perireticular cells of Mitrofanis, which form early thalamic projections, I rather refer to them simply as the 'cells of the primitive internal capsule'. What the relationship is of these cells to the striatum is not known. It is striking how the morphology and connectivity of the amphibian basal ganglionic eminence cells, as shown in Ramon y Cajal's Histologie du Système Nerveux l'Homme et des Vertébrés (Rámon y Cajal 1909–1911, Fig. 330), resembles of those cells in the embryonic internal capsule, labelled from dorsal thalamus (Molnár & Cordery 1999). The possible developmental role of these cells is not understood. They are in a position to assist the early outgrowth of thalamic projections or the sorting of the descending corticothalamic and corticospinal projections.

Rubenstein: Do these cells target the lateral ganglionic eminence (LGE) or the medial ganglionic eminence (MGE)?

Molnár: It seems that in mouse, rat and *Monodelphis domestica* the cells which possess projections to dorsal thalamus are not restricted to either the MGE or the LGE (Molnár & Cordery 1999, Molnár 1998), they reside under both. In contrast, Métin & Godement (1996) has suggested that in the embryonic hamster, these cells are restricted under the MGE, and cells with projections to the cortex are situated under the LGE. We did not observe such clear distinction (Molnár & Cordery 1999).

Hunt: How important is fasciculation for accurate pathfinding? Do chemoattractants or chemorepellants work differently depending on whether axons are single or in bundles? For example, are the fascicules easier to stop or can they penetrate developing cortex more easily than single axons?

Molnár: I suspect that at different parts of the thalamocortical pathway different extracellular milieux dominate. These differences explain why thalamic fibres show distinct pattern of fasciculation in internal capsule, striatum, and intermediate zone (Molnár et al 1998). Addressing the role of the fasciculation in fibre targeting is difficult. The results obtained in the L1 knockout mouse shows that even if larger fascicules are formed or if their order is perturbed, the thalamocortical projections still sort themselves out in the cortex (Molnár et al 1999), but this does not necessarily mean that fasciculation or the molecule L1 are not important. For instance, semaphorins are supposed to play an important role in axonal pathfinding and *in vitro* studies have begun to provide evidence for their various roles. In spite of this, the laboratories of Goodman & Shatz (Catalano et al 1998) showed that a semaphorin III knockout mouse brain looks normal. But this does not mean that semaphorin III does not have a role in axon pathfinding. I don't think that L1 or semaphorin III are redundant molecules for brain development, rather that they act in concert with others and there is a cocktail of molecules for each developmental step, so that when you start pulling them out one by one the system initially still works, but perhaps with many missing factors it will start to collapse. John Rubenstein and his colleagues (Hevner et al 1998) may be in a better position to alter pathfinding in the internal capsule because they are knocking out homeobox genes, and this may lead to the alteration of groups of the downstream genes, including ones that produce relevant cell adhesion molecules. We need to get more information about the downstream genes to *Tbr1*, *Gbx2*, *Pax6* and about the altered expression patterns in the various knockout mice. John Rubenstein has just mentioned that L1 expression is normal in these knockout mice, which rules it out as a downstream candidate molecule responsible for the abnormalities. However, in the *Tbr1*, *Gbx2* and *Pax6* knockouts no thalamocortical and corticofugal projections get through the internal capsule and they might never even get in touch with each other, therefore they do not even have a chance to start fasciculating on each other. In the L1 knockout, however, the two sets of fibres meet in the internal capsule, but then the fasciculation is abnormal. At the end, thalamic fibres reach and enter the cortex and a normal periphery related pattern is formed (Molnár et al 1999). It is yet to be determined how the disturbed fasciculation alters the topography of the initial thalamocortical targeting. An interesting experiment would be to use double labelling with DiI/DiA to examine the precision and the early path of thalamic projections. If we find that they are perturbed in the internal capsule but they sort themselves out in the cortex, this will suggest that there is room for compensation when the axons pass that region.

O'Leary: We have two findings that directly support Zoltán Molnár's results. The first is that the thalamocortical axon path through the striatum does appear to require multiple positive regulators of axon growth for proper pathfinding. In the netrin-1 knockout mouse, as thalamocortical axons pass through that part of their path, they become much more heavily fasciculated, their pathway becomes more restricted and the number of axons that make it to the cortex is reduced compared to the wild-type (Braisted et al 1999). The second is that the cells in the ventral telencephalon that can be back-filled from the dorsal thalamus before thalamocortical axons arrive are positive for Nkx2.1. This is consistent with the idea that they are part of the globus pallidus. Those cells are absent in Mash1 mutants. Correlated with their absence, we find that thalamocortical axons extend ventrally out of the dorsal thalamus and approach the hypothalamus, but they fail to turn and enter the ventral telencephalon/striatum. Instead, they just form a probst-like bundle at that point, suggesting that those cells are critical for making that pathfinding decision (Tuttle et al 1999).

Rubenstein: We did not find any abnormalities in the thalamocortical fibres in the Nkx2.1 mutant, although we have not done DiI studies, only serotonin staining. This is different to what Dennis O'Leary is describing in the Mash1 mutant.

O'Leary: It would be interesting to see if that population of cells is present or absent in the Nkx2.1 mutant.

Rubenstein: I predict it will be absent because the pallidum defects are more severe in that region than in the Mash1 mutant.

Purves: I would like to clarify the 'handshake hypothesis'. You suggested in the talk that both corticofugal and thalamic fibres know where they are going, and that there are various complicated, largely permissive interactions in the striatum. Where does this leave the original idea, namely that the two fibre systems need to interact with one another to glean this information?

Molnár: Colin Blakemore and I proposed this hypothesis in 1990 (Blakemore & Molnár 1990, Molnár & Blakemore 1990, 1991, 1995). It is based on the observation that the early corticofugal projections leave the cortex, advance in the intermediate zone and enter the internal capsule in an organized fashion synchronously with the thalamic fibres. As early corticofugal projections reach the internal capsule the front of the tips of the fibres provides a 3D representation on the cortical surface. Moreover there is a spatiotemporal difference in the arrival of these projections. When the early cortiofugal projections meet with the appropriate thalamocortical projection on a first-come, first-served basis, they become associated with one another and generate the appropriate trajectories to reach the correct region of the cortex. This is the strongest form of our hypothesis. The weakest form states that for fibre trafficking through the internal capsule the two sets of fibres have to become associated with each other. We are aware that there are many other possibilities for the thalamic fibres to sort themselves out or later modify the initial topography (see Molnár et al 2000). Catalano & Shatz (1998) showed that if the early embryonic activity in the cortex is blocked from the time of arrival of thalamocortical projections, the topography of projections can rearrange substantially. The widely held belief that the thalamocortical targeting is precise from the very beginning is based on the early postnatal tracing experiments from the cortical plate (see e.g. Agmon et al 1995), but this may not be true for embryonic stages. Naegele et al (1988) showed that in the hamster the embryonic thalamocortical projections have two to four transient side branches in relatively distinct regions of the cortical subplate, before the axon enters the cortex and reaches the cortical plate. It is possible that the stabilization of these side branches at the entry points is dependent on activity processes. This may explain why, despite of poor fasciculation, the fibres still sort themselves out in the L1 knockout (Molnár et al 1999), or change their initial layout after early cortical lesions (see discussion after Krubitzer 2000, this volume). Therefore, until we know how fasciculation relates to ordered fibre delivery we do not know whether the handshake hypothesis should be reduced to the idea that for thalamic fibres to cross through the internal capsule it is essential to associate with early corticofugal projections. There might be several levels of regulation to sort out precise thalamocortical mapping (Molnár et al 2000) and perhaps the original, early topography at the primitive internal capsule may not matter.

Levitt: You mentioned experiments on activity-dependence of developing axon trajectories by Catalano & Shatz (1998), shown by blocking sodium channel activity with TTX. Without understanding the possibility of secondary molecule changes in gene expression due to the channel block during that period of development, it is difficult to interpret those experiments. Fanny Mann, Jürgen Bolz and I, using explants from different zones of the thalamus in membranes prepared from different regions of the cortex, i.e. dorsal versus perirhinal, showed that fasciculation occurs specifically between medial thalamic and perirhinal axons, and yet we did the same experiments with more dorsal cortex in the lateral thalamus, we found that they sometimes crossed, often repelled each other, and rarely fasciculated (Mann et al 1998).

Molnár: Did you do these experiments with dissociated cells?

Levitt: No. They were explants that were separated from each other and analysed over a period of 42–72 h.

Molnár: It is important to state what surfaces are used in such fasciculation experiments because, for example, if you used membrane preparations from the selective population of presumed inhibitory cells of the internal capsule you would observe a different fasciculation pattern than when you include the whole region. Also, one might not observe fasciculation on a permissive membrane surface in culture when there is fasciculation *in vivo*. It would be important to know what are the properties of the different regions of the primitive internal capsule at embryonic day (E) 14–16. Skaliora et al (1998) showed that semaphorin G expression is localized along a stripe extending along the striatocortical junction, whereas semaphorin D is expressed in a neighbouring group of cells below the ganglionic eminence. I suspect that these distinctions might be crucial for the pausing and fasciculation of fibres in the internal capsule and perhaps they are distributed in some way in the *Tbr1*, *Gbx2* and *Pax6* knockout mice. Even if you performed your culture experiments on a membrane carpet prepared from the whole internal capsule of an E14–15 embryo, the results could not be interpreted easily. If you mixed up different subregions of internal capsule, perhaps you would lose the specific behaviour of the thalamic axons and would not observe fasciculation.

Rubenstein: Gail Martin's laboratory mutated Gbx2 (Wassarman et al 1997). We found that the thalamus differentiates abnormally and the output of axons from the thalamus is poor. We also found that the corticothalamic fibres don't reach the thalamus. This suggests that something from thalamic axons, or from the thalamus itself, is required for the cortical axons to reach the thalamus. These axons terminate in the internal capsule, which is distant from the thalamus. This suggests that a secreted molecule in the thalamus is not involved. The only caveat is that Gbx2 is expressed at low levels in the mantle of the MGE, and so far we haven't found any defects in that area. It's possible that some cells in the region of MGE are required for supporting the growth of the cortical axons to the thalamus; but if the ventral pallidum is normal, then I can't think of any other explanation besides that thalamic axons are required for cortical axons to reach the thalamus.

References

Agmon A, Yang LT, Jones EG, O'Dowd DK 1995 Topological precision in the thalamic projection to neonatal mouse barrel cortex. J Neurosci 15:549–561

Blakemore C, Molnár Z 1990 Factors involved in the establishment of specific interconnections between thalamus and cerebral cortex. Cold Spring Harbor Symp Quat Biol 55:491–504

Braisted JE, Catalano SM, Stimac R, Tessier-Lavigne M, Shatz CJ, O'Leary DDM 1999 Thalamocortical axon pathfinding is influenced by netrin-1. Soc Neurosci Abstr 25:773

Catalano SM, Shatz CJ 1998 Activity-dependent cortical target selection by thalamic axons. Science 281:559–562

Catalano SM, Messersmith EK, Goodman CS, Shatz CJ, Chedotal A 1998 Many major CNS axon projections develop normally in the absence of semaphorin III. Mol Cell Neurosci 11:173–182

Hevner RF, Miyashita E, Martin G, Rubenstein JLR 1998 Lack of thalamocortical connections in mutants affecting cortical (*TBR-1*) or thalamic (*GBX-2*) gene expression. Soc Neurosci Abstr 24:58

Krubitzer LA 2000 How does evolution build a complex brain? In: Evolutionary developmental biology of the cerebral cortex. Wiley, Chichester (Novartis Found Symp 228) p 206–226

Mann F, Zhukareva V, Pimenta A, Levitt P, Bolz J 1998 Membrane-associated molecules guide limbic and nonlimbic thalamocortical projections. J Neurosci 18:9409–9419

Métin C, Godement P 1996 The ganglionic eminence may be an intermediate target for corticofugal and thalamocortical axons. J Neurosci 16:3219–3235

Mitrofanis J 1992 Patterns of antigenic expression in the thalamic reticular nucleus of developing rats. J Comp Neurol 320:161–181

Mitrofanis J, Guillery RW 1993 New views of the thalamic reticular nucleus in the adult and developing brain. Trends Neurosci 16:240–245

Molnár Z 1998 Development of thalamocortical connections. Springer-Verlag, Berlin

Molnár Z, Blakemore C 1991 Lack of regional specificity for connections formed between thalamus and cortex in coculture. Nature 351:475–477

Molnár Z, Blakemore C 1995 How do thalamic axons find their way to the cortex? Trends Neurosci 18:389–397

Molnár Z, Cordery P 1999 Connections between cells of the internal capsule, thalamus and cerebral cortex in embryonic rat. J Comp Neurol 413:1–25

Molnár Z, Adams R, Blakemore C 1998 Mechanisms underlying the early establishment of thalamocortical connections in the rat. J Neurosci 18:5723–5745

Molnár Z, Mather NK, Katznelson A et al 1999 Disturbed fasciculation, but ordered cortical termination of thalamocortical projections in L1 knockout mice. Soc Neurosci Abstr 25:1305

Molnár Z, Higashi S, Adams R, Toyama K 2000 Earliest thalamocortical interactions. In: Kossut M (ed) The barrel cortex. FP Graham Publishing Company, London, in press

Naegele JR, Jhaveri S, Schneider GE 1988 Sharpening of topographical projections and maturation of geniculocortical axon arbors in the hamster. J Comp Neurol 277:593–607

Rámon y Cajal S 1909–1911 Histologie du Système Nerveux l'Homme et des Vertébrés, 2 vols. Maloine, Paris (trans. L. Azoulay, reprinted by Instituto Rámon y Cajal del CSIC, Madrid 1952–1955)

Skaliora I, Singer W, Betz H, Püschel AW 1998 Differential patterns of semaphorin expression in the developing rat brain. Eur J Neurosci 10:1215–1229

Tuttle R, Nakagawa Y, Johnson JE, O'Leary DDM 1999 Defects in thalamocortical axon pathfinding correlate with altered cell domains in embryonic *Mash-1* deficient mice. Development 126:1903–1916

Wassarman KM, Lewandowski M, Campbell K et al 1997 Specification of the anterior hindbrain and establishment of a normal mid/hindbrain organizer is dependent on Gbx2 function. Development 124:2923–2934

Regionalization of the cerebral cortex: developmental mechanisms and models

Pat Levitt and Kathie L. Eagleson

Department of Neurobiology, University of Pittsburgh School of Medicine, Pittsburgh, PA 15261, USA

Abstract. The cerebral cortex is composed of functionally specialized areas that have unique connections with other cortical targets and subcortical nuclei. The developmental mechanisms responsible for the formation of discrete regions must include the regulation of the expression of genes encoding proteins that control axon guidance and targeting. New data on patterns of gene expression demonstrate the early appearance of such guidance molecules, thus reflecting the early emergence of regional specification within the cortex. Transplant and cell culture studies suggest that the decisions made by neuronal progenitor cells to express region-appropriate phenotypes is controlled by the capacity of the cells to respond to and have access to specific signals. The key to understanding cortical specification may lie in determining the factors that control receptor diversity on progenitors and the temporal and spatial distribution of inductive signals within the forebrain.

2000 Evolutionary developmental biology of the cerebral cortex. Wiley, Chichester (Novartis Foundation Symposium 228) p 173–187

When discussions ensue regarding the mechanisms that control how different functional areas of the cortex arise during development and change during evolution, the focus inevitably must turn to descriptions of how relationships in cortical connectivity form and modify. We define areas of the cerebral cortex based on the appearance of the settling patterns of neurons in specific laminae, the so-called cytoarchitectonic domains. In fact, the manner in which this occurs is highly dependent upon quantitative issues of controlling cell number and phenotype through instructive mechanisms of cell proliferation and cell death (see Rakic 2000, this volume). Specialization arises from this basic regulation of cell number and diversity of phenotype. The latter is at the core of the ontogenetic events that lead to the assembly of functional circuits and the regionalization of the cortex.

Early molecular differences in axon guidance molecules

The logic in suggesting that early molecular patterning occurs in the cerebral cortex (Barbe & Levitt 1991, Levitt et al 1993), even prior to the formation of any connections, arose from an emerging understanding of the nature of how the relationship arises between the cerebral cortex and the thalamus. Specific connections form between the diencephalon and cortex prior to birth, with little evidence that would suggest there were inaccurate choices in projection patterns being made (for example, see Wise & Jones 1977, De Carlos & O'Leary 1992, Erzurumulu & Jhaveri 1992). We have made an assumption that general mechanisms of axon pathfinding and targeting are conserved among species and throughout the nervous system. There is some logic, therefore, in drawing parallels between the identification of discrete expression patterns of guidance molecules among different populations of embryonic neurons both in invertebrates and in many subcortical regions of the vertebrate nervous system (Goodman 1996), and those same families of molecules within the early developing cerebral cortex. We suggested such regional specification occurred following the discovery of the early, restricted expression of the limbic system-associated membrane protein (LAMP) in medial prefrontal, insular and perirhinal areas of the cerebral cortex, and in corresponding anterior and medial thalamic nuclei (Levitt 1984, Horton & Levitt 1988, Zacco et al 1990). Gene cloning identified LAMP as a member of the Ig superfamily of cell adhesion molecules, and functional cell biological studies indicated that LAMP can serve as an axon guidance molecule (Keller et al 1989, Pimenta et al 1995, Zhukareva & Levitt 1995, Mann et al 1998). It became apparent that there might be an early commitment of cells to express LAMP soon after differentiating into neurons. This was based on our observation that LAMP expression could be detected first on migrating neurons, prior to their arrival in the cortical plate. Subsequent transplant studies, in which we found that the LAMP phenotype was immutable once neuronal differentiation occurred, even when tissue was placed in ectopic locations (Barbe & Levitt 1991), further supported the notion of an early molecular patterning of the cerebral cortex. With just one example, however, the findings conceivably could reflect a special, evolutionarily conserved mechanism for ensuring that allo- and mesocortical regions are maintained phylogenetically (O'Leary 1989, O'Leary et al 1994, Levitt et al 1993, 1997).

The geography of axon guidance molecules in the cerebral cortex recently has become far more complex, with two other axon guidance families being shown to have early and restricted expression patterns. Members of the eph family of tyrosine kinase receptors and complementary ephrin ligands are distributed in a highly complex, heterogeneous fashion (Gao et al 1998, Mackarehtschian et al 1999); ephrin A5 initially is expressed most heavily in presumptive somatosensory cortex at embryonic day (E) 17 in the rat and the complementary ligand that induces repulsive axon guidance behaviour, ephA5, is found most prominently in presumptive frontal cortex. Members of the class II cadherin family of adhesion molecules, cad6, 8 and 11, exhibit differences in expression as early as E14.5 in the mouse cortex (Redies & Takeichi 1996).

The beginnings of discrete transcription factor expression in the cerebral wall

If one adds to this discussion other examples of regionally distributed proteins (without axon guidance activity; Arimatsu et al 1992, 1994, Cohen-Tannoudji et al 1994, Snyder et al 1998), parcellation of the entire cerebral cortex at the molecular level now appears more the norm than the exception. What could account for the early differences in gene expression patterns in the cortex? As in other regions of the neuraxis, it is likely that in the cerebral wall there will be complex combinations of transcription factors that control gene expression among different populations of progenitor cells. Although the initial prosomere model of forebrain development had not at the time included examples of such transcription factor patterns in the cerebral wall (Puelles & Rubenstein 1993, Rubenstein et al 1994), recent gene hunting has discovered some interesting, early specializations. Emx1 and Emx2 are expressed throughout the cerebral wall, but in gradients between anterior and posterior domains (Briata et al 1996, Mallamaci et al 1998). The cortical hem in the ventral midline is a site of complex transcription factor expression, including Wnt and Gli3, both of which may contribute to the development of hippocampal fields (Grove et al 1998). Our laboratory recently has used subtraction–differential expression methods to discover other unique transcription factor patterns in the early developing cerebral wall (D. Campbell & P. Levitt, unpublished results 1999). As this area of investigation progresses, we would predict the emergence of more and more complex expression patterns of transcription factors, each of which will need to be tested directly for their role in early cortical patterning. A variety of approaches already are being used to show the potency of transcription factors in controlling the development of many different forebrain regions (see this volume: Rubenstein 2000, Boncinelli et al 2000, Parnavelas et al 2000). One assumes that the potency of the complex interactions of transcription factors that control gene expression results from the regulation of downstream effectors that directly modulate the generation of specific neuronal populations (Edlund & Jessell 1999).

Cell location is critical for modulating developmental potential

In our early studies of LAMP expression, we were able to show, using embryonic tissue transplants, that the movement of progenitor cells from regions of the cerebral wall that normally do not give rise to LAMP$^+$ cells (for example, sensorimotor) to a region that normally does produce LAMP$^+$ neurons results in the induction of LAMP expression by the transplanted neurons (Barbe & Levitt 1991). In fact, the alteration in LAMP phenotype is paralleled by changes in thalamocortical and corticocortical projections that are identical to normal LAMP$^+$ perirhinal and insular cortex (Barbe & Levitt 1992, 1995). Thus, in the context of two strict phenotypic parameters, the expression of a specific axon guidance molecule and patterns of axon projections, we were able to change the fate of non-limbic cortical neurons, such that they express a 'limbic' phenotype. The *in vivo* experiments suggest that progenitor cells from any region of the developing cerebral wall have the potential to express these 'limbic' traits. There must be some control, therefore, imposed by the location of the progenitors when they differentiate into neurons. This suggests the possibility that there are restricted, organizing domains in the forebrain that may be responsible for producing diffusible signals that regulate early regional features of the cerebral cortex.

The role of growth factor signalling in regulating molecular specification of the cortex

The powerful actions of transcription factors, expressed in restricted anatomical domains of the neuraxis, on downstream genes that encode proteins controlling cell fate have been elucidated both in *Drosophila* and vertebrate nervous systems (Edlund & Jessell 1999). Focusing on the forebrain, for example, the combined mutation of *Otx1* and *Otx2* results in a dramatic shift anteriorly in the expression of fibroblast growth factor (FGF)8 (Acampora et al 1997, Suda et al 1997), a secreted growth factor responsible for establishing the midbrain–hindbrain boundary. The mutation results in the posteriorization of the diencephalon. Mutations of the transcription factors *Nkx2.1* and *Gbx2* result in the loss of the medial ganglionic eminence and the dorsal thalamus, respectively (Rubenstein 2000, this volume), and a major reorganization of the telencephalon in each mutation. Mutation of *Nkx2.1* results in a loss of *sonic hedgehog* expression in the rostral forebrain. These and other mutations document the rather direct manner in which alterations to key genes can lead to complex reorganization of nervous system architecture.

Our laboratory has maintained the goal of identifying as many of the elements in the signal cascades that are responsible for early cortical patterning. We have focused on defining the signals directly upstream of controlling the expression of axon guidance molecules, because these latter proteins are essential for defining the functional features of the cortex. Using LAMP expression as an assay for a specific regional feature of cortical neurons, we demonstrated that progenitors isolated from different domains of the cerebral wall, prior to neuron production, will differentiate *in vitro* according to their original location (Ferri & Levitt 1993). The fact that we were able to modify sensorimotor progenitors in the transplant studies, however, suggested that signals are present in limbic regions that effectively control LAMP expression *in vivo*. Our recent efforts have led to the discovery that the erbB receptor family is involved in the signalling cascade that regulates LAMP expression (Ferri & Levitt 1995, Eagleson et al 1998). Moreover, we found a curious dependency of LAMP induction, through erbB receptor signalling, on a specific extracellular matrix component type IV collagen. This protein is expressed transiently in the early ventricular zone of the cerebral wall during the time of neurogenesis (Eagleson et al 1996). In our studies, progenitors harvested from presumptive sensorimotor or visual domains were equally responsive to the erbB receptor ligands transforming growth factor (TGF)α or β-heregulin (Eagleson et al 1998). In all instances, however, exposure to the ligand was required when the cells were actively in the cell cycle (Ferri et al 1996, Eagleson et al 1997, 1998). A delay of just a few days in adding the ligand, when proliferation in the cultures ceased and neuronal differentiation occurred, resulted in a complete absence of LAMP induction.

Our most recent analysis revealed another important characteristic of progenitor cells from the cerebral wall. Whereas 70–80% of the neurons that differentiated from sensorimotor or visual progenitors expressed LAMP when exposed to TGFα, only about 50% of the neurons expressed LAMP following β-heregulin exposure (Eagleson et al 1998). Under all conditions, the same number of neurons differentiated and survived, indicating that the data reflect actual differences in the responsiveness of the progenitors to the inductive signals. Each ligand acts through different members of the erbB receptor family, suggesting to us the possibility that progenitor cells in the cerebral wall, early in embryonic development, are heterogeneous with respect to their receptor repertoire. This proved to be correct; double-label immunocytochemical analysis revealed complex combinations of erbB receptor expression by early progenitors (Eagleson et al 1998, Eagleson & Levitt 1999). These findings have formidable implications, because it is likely that the capacity for responsiveness to a particular ligand may be distinct amongst populations of progenitors in the cerebral ventricular zone.

The fact that progenitor cells in the early cerebral wall are heterogeneous is highlighted further by recent studies of Arimatsu and co-workers (Arimatsu et al 1999). Using a phenotypic marker (latexin) for a subpopulation of projection neurons in lateral cortex, their experiments showed that early in development, by E13, the dorsal and lateral domains of the cerebral wall contain progenitors exhibiting a pronounced difference in their capacity to respond to environmental signals that induce latexin.

Towards a model of cortical regionalization

Overwhelming data now indicate that the cerebral cortex emerges from an early regionalization, which occurs prior to subcortical input. The regionalization effectively parcellates domains that subsequently develop into areas that underlie functional specialization. In this context, perhaps the most remarkable result reported at this symposium was the apparently normal regionalization of the cerebral cortex in the absence of a dorsal thalamus in the *Gbx* knockout mouse (Rubenstein 2000, this volume). It is the definition of the molecular mechanisms that control thalamic and cortical patterning that may provide insights into evolutionary change in the cerebral cortex. In our own work, we have emphasized the importance of understanding the capacity of cells to respond to regulatory signals, and the role of cell position in the developing cortical 'field' in defining the response properties of progenitors. Position of cells in the field may regulate several aspects of development. First, it is likely that regionalizing signals that impact on progenitors in the pallium are restricted spatially. Thus, only those progenitors that reside in proximity to the source of the signal will respond (Fig. 1).

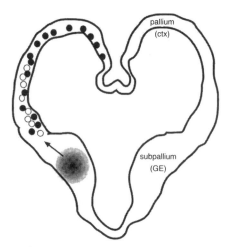

FIG. 1.   The diagram depicts a model of induction of regional phenotypes in the cerebral cortex. Drawing represents the forebrain of a rodent at a time equivalent to the onset of neurogenesis in the cerebral wall. The large, darkened region in the subpallium represents a source of an inductive signal that will diffuse over a relatively restricted distance. The small black circles represent progenitors that express the receptors required to respond to the inductive signal. Note that only those cells in close proximity to the source of the signal will respond, but all progenitors throughout the lateromedial extent of the cerebral wall could respond if given the opportunity. The white circles represent a subpopulation of progenitors that have a restricted distribution in the cerebral wall. Compared to other progenitors in the cerebral wall, these cells have unique receptor features that control a different, more restricted response to an inductive signal. ctx, cortex; GE, ganglionic eminence.

Second, cell position itself may control the expression of receptors that even allow progenitors to respond to regionalizing signals. The combination of response complexity and signal accessibility is a common theme in the regulation of cellular diversity in the nervous system (Lillien 1998).

Concluding remarks

It is clear that relatively simple modifications in gene expression can result in profound changes in cortical organization. For example, in our early transplant studies, when we generated cortical fields with mixed LAMP$^+$ and LAMP$^-$ neurons, we obtained animals with mixed thalamic projections from limbic and ventrobasal nuclei (Barbe & Levitt 1992, Levitt et al 1997). In fact, this kind of projection pattern normally never develops, and suggests that early changes in the signals that control the expression of axon guidance molecules can produce new cortical-subcortical relationships that are at the core of functional change. The challenge will be to understand how molecular changes are regulated both developmentally and phylogenetically.

Acknowledgements

The results from our laboratory that were reported in this chapter were supported by National Institute of Medical Heath Grant MH45507 (P.L.) and March of Dimes Basic Research Grant 6-FY97-0352 (P.L. and K.E.).

References

Acampora D, Avantaggiato V, Tuorto F, Simeone A 1997 Genetic control of brain morphogenesis through *Otx* gene dosage requirement. Development 124:3639–3650

Arimatsu Y, Miyamoto M, Nihonmatsu I et al 1992 Early regional specification for a molecular neuronal phenotype in the rat neocortex. Proc Natl Acad Sci USA 89:8879–8883

Arimatsu Y, Nihonmatsu I, Hirata K, Takiguchi-Hayashi K 1994 Cogeneration of neurons with a unique molecular phenotype in layers V and VI of widespread lateral neocortical areas in the rat. J Neurosci 14:2020–2031

Arimatsu Y, Ishida M, Takiguchi-Hayashi K, Uratani Y 1999 Cerebral cortical specification by early potential restriction of progenitor cells and later phenotype control of postmitotic neurons. Development 126:629–638

Barbe MF, Levitt P 1991 The early commitment of fetal neurons to the limbic cortex. J Neurosci 11:519–533

Barbe MF, Levitt P 1992 Attraction of specific thalamic input by cerebral grafts depends on the molecular identity of the implant. Proc Natl Acad Sci USA 89:3706–3710

Barbe MF, Levitt P 1995 Age-dependent specification of the corticocortical connections of cerebral grafts. J Neurosci 15:1819–1834

Boncinelli E, Mallamaci A, Muzio L 2000 Genetic control of regional identity in the developing vertebrate forebrain. In: Evolutionary developmental biology of the cerebral cortex. Wiley, Chichester (Novartis Found Symp 228) p 53–66

Briata P, Di Blas E, Gulisano M et al 1996 EMX1 homeoprotein is expressed in cell nuclei of the developing cerebral cortex and in axons of the olfactory sensory neurons. Mech Dev 57: 169–180

Cohen-Tannoudji M, Babinet C, Wassef M 1994 Early determination of a mouse somatosensory cortex marker. Nature 368:460–463

De Carlos JA, O'Leary DDM 1992 Growth and targeting of subplate axons and establishment of major cortical pathways. J Neurosci 12:1194–1211

Eagleson KL, Levitt P 1999 Complex signaling responsible for molecular regionalization of the cerebral cortex. Cereb Cortex 9:562–568

Eagleson KL, Ferri RT, Levitt P 1996 Complementary distribution of collagen type IV and the epidermal growth factor receptor in the rat embryonic telencephalon. Cereb Cortex 6:540–549

Eagleson KL, Lillien L, Chan AV, Levitt P 1997 Mechanisms specifying area fate in cortex include cell-cycle-dependent decisions and the capacity of progenitors to express phenotype memory. Development 124:1623–1630

Eagleson KL, Daigneau L, Levitt P 1998 The role of erbB receptor signaling in cell fate decisions by cortical progenitors: evidence for a biased, lineage-based responsiveness to different ligands. Mol Cell Neurosci 12:349–362

Edlund T, Jessell TM 1999 Progression from extrinsic to intrinsic signaling in cell fate specification: a view from the nervous system. Cell 96:211–224

Erzurumlu RS, Jhaveri S 1992 Emergence of connectivity in the embryonic rat parietal cortex. Cereb Cortex 2:336–352

Ferri RT, Levitt P 1993 Cerebral cortical progenitors are fated to produce region-specific neuronal populations. Cereb Cortex 3:187–198

Ferri RT, Levitt P 1995 Regulation of regional differences in the fate of cerebral cortical neurons by EGF family-matrix interactions. Development 121:1151–1160

Ferri RT, Eagleson KL, Levitt P 1996 Environmental signals influence expression of a cortical areal phenotype in vitro independent of effects on progenitor cell proliferation. Dev Biol 175:184–190

Gao PP, Yue Y, Zhang JH, Cerretti DP, Levitt P, Zhou R 1998 Regulation of thalamic neurite outgrowth by the Eph ligand ephrin-A5: implications in the development of thalamocortical projections. Proc Natl Acad Sci USA 95:5329–5334

Goodman CS 1996 Mechanisms and molecules that control growth cone guidance. Annu Rev Neurosci 19:341–377

Grove EA, Tole S, Limon J, Yip L, Ragsdale CW 1998 The hem of the embryonic cerebral cortex is defined by the expression of multiple Wnt genes and is compromised in Gli3-deficient mice. Development 125:2315–2325

Horton HL, Levitt P 1988 A unique membrane protein is expressed on early developing limbic system axons and cortical targets. J Neurosci 8:4653–4661

Keller F, Rimvall K, Barbe MF, Levitt P 1989 A membrane glycoprotein associated with the limbic system mediates the formation of the septo-hippocampal pathway in vitro. Neuron 3:551–561

Levitt P 1984 A monoclonal antibody to limbic system neurons. Science 223:299–301

Levitt P, Ferri RT, Barbe MF 1993 Progressive acquisition of cortical phenotypes as a mechanism for specifying the developing cerebral cortex. Perspect Dev Neurobiol 1:65–74

Levitt P, Barbe MF, Eagleson KL 1997 Patterning and specification of the cerebral cortex. Annu Rev Neurosci 20:1–24

Lillien L 1998 Neural progenitors and stem cells: mechanisms of progenitor heterogeneity. Curr Opin Neurobiol 8:37–44

Mackarehtschian K, Lau CK, Caras I, McConnell SK 1999 Regional differences in the cortical subplate revealed by ephrin-A5 expression. Cereb Cortex 9:601–610

Mallamaci A, Iannone R, Briata P et al 1998 EMX2 protein in the developing mouse brain and olfactory area. Mech Dev 77:165–172

Mann F, Zhukareva V, Pimenta A, Levitt P, Bolz J 1998 Membrane-associated molecules guide limbic and nonlimbic thalamocortical projections. J Neurosci 18:9409–9419

O'Leary DDM 1989 Do cortical areas emerge from a protocortex? Trends Neurosci 12:400–406

O'Leary DDM, Schlaggar BL, Tuttle R 1994 Specification of neocortical areas and thalamocortical connections. Annu Rev Neurosci 17:419–439

Parnavelas JG, Anderson SA, Lavdas AA, Grigoriou M, Panchis V, Rubenstein JLR 2000 The contribution of the ganglionic eminence to the cell types of the cerebral cortex. In: Evolutionary developmental biology of the cerebral cortex. Wiley, Chichester (Novartis Found Symp 228) p 129–147

Pimenta A, Zhukareva V, Barbe MF et al 1995 The limbic system-associated membrane protein is an Ig superfamily member that mediates selective neuronal growth and axon targeting. Neuron 15:287–297

Puelles L, Rubenstein JL 1993 Expression patterns of homeobox and other putative regulatory genes in the embryonic mouse forebrain suggest a neuromeric organisation. Trends Neurosci 16:472–479

Rakic P 2000 Radial unit hypothesis of neocortical expansion. In: Evolutionary developmental biology of the cerebral cortex. Wiley, Chichester (Novartis Found Symp 228) p 30–45

Redies C, Takeichi M 1996 Cadherins in the developing central nervous system: an adhesive code for segmental and functional subdivisions. Dev Biol 180:413–423

Rubenstein JLR 2000 Intrinsic and extrinsic control of cortical development. In: Evolutionary developmental biology of the cerebral cortex. Wiley, Chichester (Novartis Found Symp 228) p 67–82

Rubenstein JLR, Martinez S, Shimamura K, Puelles L 1994 The embryonic vertebrate forebrain: the prosomeric model. Science 266:578–580

Snyder SE, Pintar JE, Salton SRJ 1998 Developmental expression of VGF mRNA in the prenatal and postnatal rat. J Comp Neurol 394:64–90

Suda Y, Matsuo I, Aizawa S 1997 Cooperation between Otx1 and Otx2 genes in developmental patterning of rostral brain. Mech Dev 69:125–141

Wise SP, Jones EG 1977 Cells of origin and terminal distribution of descending projections of rat somatic sensory cortex. J Comp Neurol 175:129–157

Zacco A, Cooper V, Chantler PD, Hyland-Fisher S, Horton HL, Levitt P 1990 Isolation, biochemical characterization and ultrastructural localization of the limbic system-associated membrane protein (LAMP), a protein expressed on neurons comprising functional neural circuits. J Neurosci 10:73–90

Zhukareva V, Levitt P 1995 The limbic system-associated membrane protein (LAMP) selectively mediates interactions with specific central neuron populations. Development 121:1161–1172

DISCUSSION

Karten: You showed how it is possible to alter specific cell types. However, from an evolutionarily perspective, what stands out is how constant the different cell types are.

Papalopulu: But you also showed that it is possible to change the connections, which suggests that it is not that difficult to re-wire a brain — only small changes are required.

Levitt: Yes, and there are other ways of re-wiring the brain other than transplantation. For example, it is possible to get substantia nigra neurons to project into the iris. If you put the iris in the middle of the medial forebrain bundle, the axons from the nigra grow in. They look like sympathetic axons, and yet no one would suggest that the nigra neurons have turned into sympathetic neurons. There are many ways to alter connectivity, but developmentally it is not that complicated. If you do the experiment at the right time, those transplants have connectivity unlike any I have seen in any normal or homotypic transplant combination. We should re-do those experiments now that we have many more molecular markers.

Papalopulu: Doesn't this also imply that the analysis of these connectivity patterns cannot give us any information about which parts of the brain are homologous between species?

Levitt: I'm not sure. The two questions we are trying to answer are: what is homologous to what; and how do these connections change over time? We found that it is possible to change the organization of projections developmentally.

Rubenstein: It's dangerous to use connectivity to look at evolutionary homologies. We should be careful to avoid using secondary, tertiary and quaternary steps in development to homologize evolutionary similarities, although if they are conserved, it does give additional weight to homology arguments.

Karten: At the level of the brainstem, there is little dispute about the stability of many of the connections. We are arguing about what's going on in the forebrain, and there are local levels at which there are uncertainties, but there are many places where the connections are extremely stable across evolution. We don't know how these connections are stabilized across evolution.

Rakic: Transplants are a useful experimental tool, but I would just like to point out these experiments demonstrate the capacity for plasticity, but the existence of plasticity doesn't negate specificity.

Levitt: The point is not whether or not there is specificity. The point is that part of the specificity relates to where you are within the developing field. The experimental manipulations are designed to change the time and the location where the cells develop. The experiment in which we added growth factor receptors (Burrows et al 1997) was to change time. We could not think of another way to speed up the maturation of the ventricular zone progenitors.

Rubenstein: I liked your experiments, but I would like to qualify that you are talking not about the entire subventricular zone, but the subventricular zone of the pallium, because the subpallial subventricular zone is a robust producer of neurons.

Levitt: Yes. I would like to add that we have also analysed what happens in the subpallium after adding extra epidermal growth factors (EGF) receptors. We found that astrocytes are generated prematurely. It's not a minor effect; there is about an eightfold difference compared to normal. What's really interesting is that in the pallium, astrocytes initially distribute in a bilaminar pattern — up in the marginal zone and down in the deep cortical plate — and then they fill the space in-between. In the subpallium, however, that migratory pattern is not present. The astrocytes migrate more uniformly. We see the same phenotypic shift, i.e. a cell that normally makes a neuron, now makes an astrocyte, but the migratory properties are location-specific (Burrows et al 2000).

Karten: I would like to ask what is the general consensus on the suggestion that connections are highly changeable? How constant are they in terms of circuits, because the end product is the behaviour, and similar circuits often mediate similar behavioural traits.

Hodos: In my opinion, they are relatively constant. The behavioural experiments also show this. For example, the results of lesion experiments in birds and reptiles, which are mostly visual experiments but also include auditory and somatosensory experiments, suggest a high degree of constancy.

Butler: I would say that most connections are constant, but there are dramatic instances of changes in connections. For example, in amphibians the collothalamus, which receives midbrain roof input, projects predominantly to the striatum, and so there was a large shift of those projections up to the pallium with the origin of amniotes. In order to look at homologies, it is necessary to construct total evidence trees. You have to consider connections, location, histochemistry and gene expression studies, and then you have to come to the most parsimonious conclusion based on all these lines of evidence.

Pettigrew: Albert et al (1998) did some elegant mathematical analyses of electric fish, for which there's abundant molecular neurological and morphological data. They compared the trees and were able to distinguish areas of agreement and conflict. There are a couple of cases where the neurological characters give the wrong answer, but the congruence of the neurological data is high.

Hunt: I would like to say that I disagree with Bill Hodos' comment about behaviour being the same. The bird forebrain, for example, could be: striatum, which has largely been dismissed; cortex, which would fit with Bill's idea; or amygdala. If you imagine that the forebrain of birds is largely amygdala, then the behaviour becomes a lot more interesting. In mammals, the amygdala is important for picking out what's important in the environment, assigning salience to a stimulus and in 'emotional learning'. We might therefore ask what capacities do mammals possess that birds lack, and vice versa, that reflect these different routes of forebrain evolution.

Hodos: If you look at the kinds of behavioural effects generated from lesions in the striatum, you find principally sensory types of deficits, e.g. changes in psychophysical thresholds, whereas for lesions in the amygdala, you would expect to see more in the way of cognitive effects.

Karten: We are also starting to collect data on the coding of neurons within the tectofugal system. The preliminary results look identical to the results from the inferior and posterior temporal lobe, in terms of motion detection, lead response properties and the kinetics of responses, i.e. global responses to small objects moving at high velocity.

Hunt: What about imprinting? This is characteristic of birds.

Karten: Most of the areas that have been associated with imprinting are not in those areas we've been arguing about; they are mostly Emx negative and in the hyperstriatum ventrale. On the one hand, we're trying to find out which areas in non-mammals might have some of the cells that we find in mammals in order to track the phylogeny of the cells. On the other hand, what are the unique properties of birds? In other words, we don't expect identity. The questions are, what are the similarities, what are the homologies, what are the transformations and what are the unique properties? We haven't really begun to address this.

Kaas: I have a brief comment on the issue of changing connection sites. It looks like the general pattern of projections from ventral posterior region of the cortex is to both S1 and the lateral areas, i.e. S2 and the parietal ventral area, PV. But it also looks like higher primates have dropped the parallel projection to the lateral cortex, and instead the lateral cortex is activated serially from primary cortex. This is an example of a changing connection. There are advantages in both parallel and serial processing, so it is interesting to think about how such changes might impact on behaviour.

Herrup: I would like to change the subject and ask whether all cells in the limbic system-associated membrane protein (LAMP)-positive region are LAMP-positive?

Levitt: I would say that at least 70% of the cells are positive in any particular area. We have to do serial sections through them, and with a membrane protein this is a little difficult. No glial cells express LAMP. Most of the proteins I mentioned are nervous system-specific, and are not expressed in somites or other tissues. We haven't yet focused on doing double labelling for γ-aminobutyric acid (GABA) or glutamate.

Rubenstein: I like the idea that LAMP itself is enough to change connectivity, so perhaps if you made an enhancer mutation in the *LAMP* gene to give rise to ectopic expression, you may see connectivity changes.

Puelles: Can you say anything about LAMP expression in the chick?

Karten: It stains heavily in the rotundus. We then looked for it in mice and rats, but we could not find any evidence for it. We did see comparable staining patterns in the cerebellum and a few other places. I was just thinking that it may be time to go back and look in the squirrel, where we now know where the homologous cell group is.

Puelles: Was it expressed in the posterior complex in the thalamus?

Karten: Only at low levels. The limbic areas of the hypothalamus were also positive, but less so than the rotundus. We did not look at the limbic areas of the telencephalon in the chick because we didn't know where they were. I was hoping that it would also be present at high levels in the hippocampus. The antibody stained something in that region, but the staining wasn't robust, as it was in the cerebellum and rotundus.

Reiner: We also found that the antibody stained the neostriatum and in the hyperstriatum ventrale in pigeons.

Levitt: In primate striatum, LAMP is expressed in the medial striatum, nucleus accumbens and striatal patch regions (Côté et al 1995).

Puelles: Did you look at the expression patterns in the lateral amygdala?

Levitt: In collaboration with Andre Parent's laboratory, we have looked at the expression patterns in the amygdala. It has a complicated distribution pattern in the amygdala, including heavy labelling in a region called the extended amygdala (Côté et al 1996).

Karten: But you also saw staining in a number of cortical areas, it was just heavier as I recall.

Levitt: As you go from lisencephalic to gyrencephalic animals, the expression becomes much sharper and more restricted. It is possible that this parallels the precision of the projections that occur within gyrencephalic animals.

Rubenstein: I would like to ask if the expression patterns of these transforming growth factor (TGF)α ligands and Erb receptors are consistent with their potential role in regionalization, and how do the analyses of the mutations of these genes fit with this hypothesis?

Levitt: The TGFα mapping has been done by *in situ* hybridization, but it is a bit risky to draw conclusions regarding activity zones because it is a secreted factor. Marianne Blum's laboratory (Lazar & Blum 1992) and Harley Kornblum's laboratory (Kornblum et al 1997) have done most of this. The highest expression levels of TGFα transcript are in the subpallium. We have also found that the cortical regions closest to the subpallium, looking in both the sagittal and coronal planes, express the *lamp* gene. The VGF transcript and protein also have a restricted expression pattern (Snyder et al 1998). It is present only in lateral and medial cortical domains, and my unpublished work with Kathie Eagleson (1999) shows that it is regulated by NT3 and brain-derived neurotrophic factor (BDNF). Careful mapping of the ligands themselves, and not the transcripts, has not been done. The third set are the β-heuregulins. Marchionni et al (1993) have published some images showing a number of the spliced variants with restricted patterns of expression in the pallium and subpallium. It turns out that one of the spliced variants has a restricted expression in the lateral domains of the cortex, but again, ligand mapping has not been done.

Reiner: I don't think anyone would argue that projections cannot change throughout evolution. However, it is dangerous to place one line of evidence above others. For example, in his presentation, Pat Levitt said that LAMP was expressed in chick dorsal root ganglia (DRGs) and not mammalian DRGs, but no one would say that the DRGs are not homologous because they express different genes. I would say that we have to look at the weight of evidence, and I personally don't know which evidence stands above all other kinds of evidence.

Rubenstein: Transcription factors that control cell fate and control regional identity may be better markers of regional identity and cell fate, then genes encoding cell surface proteins, particularly if their regional and temporal expression is conserved in embryos from diverse vertebrate species. It is likely that these transcription factors regulate the expression of cadherins and the Ig superfamily, for example. Species-specific mutations in the enhancers of the cell surface proteins can alter their expression patterns, which in turn can have profound effects on conductivity.

Karten: But not all cells in any given region express the same transcription factor.

Rubenstein: It depends what you're talking about. If you examine expression in the ventricular zone, for instance, you will obtain more precise answers than if you look at later stages once cells begin to migrate away from their place of birth. Much of the confusion about the spread of transcription factor expression beyond 'compartment' boundaries may be a result of tangential migrations that are part of later developmental programmes.

Reiner: It seems to me that one risk of using transcription factors as identifiers of identity, is that you may mistake a behaviour imparted by that transcription factor for an indicator of identity.

Herrup: On occasion the same gene can serve two different functions and can join a different network of genes. Therefore, we need to specify development times carefully when analysing these factors.

Another point is that I keep thinking about Lewis Wolpert's description of morphogen gradients in developmental fields. When Pat Levitt spoke about the sharpness of the LAMP gradient in limbic cortices, I started to think about the role of size in generating these gradients. I wonder whether size alone can not only sharpen the gradients, but also give the opportunity to create new transcription factor patterns in regions where the gradients become ambiguous because there's too great a distance between the two ends of the gradient.

O'Leary: I can't speak about transcription factors, but a potentially relevant example might be the action of graded distributions of axon guidance molecules. For example, it is reasonable to assume that a given axon guidance molecule expressed in a graded distribution can only govern axon guidance precisely over a specific distance. This distance would be dictated by several factors, including the effective concentration range of the guidance molecule that can influence growth cones, and the change in concentration over the length of the growth cone required to affect growth cones. For example, Bonhoeffer's work, indicates that a growth cone can respond to a concentration change of roughly of 1 or 2% over the length of the growth cone, and below this level it will not respond (Rosentreter et al 1998). This puts a limitation on the distance that a given graded guidance molecule can effectively guide axons. Geoff Goodhill has estimated that this distance is roughly 5–10 mm. If a structure exceeds that size, additional guidance systems would likely be required.

References

Albert JS, Lannoo MJ, Yuri T 1998 Testing hypotheses of neural evolution in gymnotiform electric fishes using phylogenetic character data. Evolution 52:1760–1780

Burrows RC, Wancio D, Levitt P, Lillien L 1997 Response diversity and the timing of progenitor cell maturation are regulated by developmental changes in EGFR expression in the cortex. Neuron 19:251–267

Burrows RC, Lillien L, Levitt P 2000 Mechanisms of progenitor cell maturation are conserved in the striatum and cortex. Dev Neurosci 22:7–15

Côté PY, Levitt P, Parent A 1995 Distribution of limbic system-associated membrane protein immunoreactivity in primate basal ganglia. Neuroscience 69:71–81

Côté PY, Levitt P, Parent A 1996 Limbic system-associated membrane protein (LAMP) in primate amygdala and hippocampus. Hippocampus 6:483–494

Kornblum HI, Hussain RJ, Bronstein JM, Gall CM, Lee DC, Seroogy KB 1997 Prenatal ontogeny of the epidermal growth factor receptor and its ligand, transforming growth factor α, in the rat brain. J Comp Neurol 380:243–261

Lazar LM, Blum M 1992 Regional distribution and developmental expression of epidermal growth factor and transforming growth factor-α mRNA in mouse brain by a quantitative nuclease protection assay. J Neurosci 12:1688–1697

Marchionni MA, Goodearl ADJ, Chen MS et al 1993 Glial growth factors are alternatively spliced erbB2 ligands expressed in the nervous system. Nature 362:312–318

Rosentreter SM, Davenport RW, Löschinger J, Huf J, Jung J, Bonhoeffer F 1998 Response of retinal ganglion cell axons to striped linear gradients of repellent guidance molecules. J Neurobiol 37:541–562

Snyder SE, Pintar JE, Salton SR 1998 Developmental expression of VGF mRNA in the prenatal and postnatal rat. J Comp Neurol 394:64–90

Organizing principles of sensory representations

Jon H. Kaas

Department of Psychology, Vanderbilt University, 301 Wilson Hall, Nashville, TN 37240, USA

Abstract. Mammalian brains vary greatly in size and expanse of neocortex. Yet, regardless of the extent, much of the cortex consists of orderly representations of receptor surfaces. Many of these representations closely reflect the order of their receptor sheet. The fidelity of the match can be so exact that discontinuities in the receptor sheet, such as the optic disc of the retina or the separations between fingers, are reflected in the representations. If parts of the receptor surface are duplicated or missing in development, representations are appropriately altered. Such isomorphisms suggest that the receptor sheet instructs the central representations to influence the course of their development. Representations may be fractured into a mosaic of small partial maps, possibly as a result of competing factors in development. Parts of receptor surfaces can achieve proportionately more than their share of a sensory representation. The congruence of borders between representations suggests the transfer of instructions across borders. Neural activity patterns are a likely source of developmental information in all these instances.

2000 Evolutionary developmental biology of the cerebral cortex. Wiley, Chichester (Novartis Foundation Symposium 228) p 188–205

The brains of mammals are highly variable. The most obvious variation is in size. The surface area of one cerebral hemisphere in the smallest mammals with the smallest brains, for example, is on the order of 1.5 cm^2 (Catania et al 1999), while the surface area in humans is about 800 cm^2 (Van Essen & Drury 1997), a more than 500-fold increase. The scaling problems posed by such great differences in size seem to be solved in part by differences in the numbers of areas and modules within brains (Kaas 1988, 1993). Mammals with small brains seem to devote about half or more of neocortex to a small number (10 or less) of sensory and motor representations. Mammals with large brains also devote much of neocortex to sensory and motor representations. To some extent, larger brains have larger representations, especially the shared 'primary' representations, but they also have more representations. Macaque monkeys, for example, have over 30 visual areas. The larger sensory representations of larger brains may also be more modular, as a solution to the problem of maintaining connections in larger areas. Sensory representations also differ within and across species in the degree and types of laminar differentiation.

While the variation across mammals in brain size and organization is impressive, the sensory areas in these brains express a number of features in common. These common features presumably are maintained in evolution by their utility or by biological constraints on brain development. Here we consider some of the features of sensory representations commonly expressed, and then functional and developmental implications.

Sensory representations reflect the order of the receptor sheet

Visual, auditory or somatosensory representations, at least many of them, are obviously retinotopic, cochleotopic or somatotopic (Merzenich & Kaas 1980). Often, large parts of the sensory sheet are continuously represented. Somatosensory representations, for example, may be continuous, although folded, for locations activated by stimuli progressing from tail to tongue. Representations may be of only part of a continuous receptor surface. The contralateral half of the body surface is typically represented in somatosensory areas and nuclei, and the contralateral visual hemifield and corresponding parts of the retina of each eye are typically represented in visual areas and nuclei, but the representations are at least grossly continuous. When more of the sensory sheet is included in a representation, such as when ipsilateral parts of the mouth are included in somatosensory representations (Johnson 1990) or parts or all of the ipsilateral hemifield are included in visual representations, as in the superior colliculus (Kaas et al 1973a) or visual cortex of Siamese and albino cats (Kaas & Guillery 1973, Hubel & Wiesel 1971), the extended representations remain topographic.

Disruptions in the receptor sheet
are often reflected in sensory representations

While the organizations of sensory maps are often explored with recording electrodes, these maps have an anatomical substrate that often can be made visible. Surface anatomical features such as fissures and dimples have long been known to distinguish the representational boundaries of digits of the forepaw and other body parts in the cortex of racoons and other mammals (see Johnson 1990). As a better known example, the existence of a separate structure, the so-called barrel, for each whisker of the face has been clearly demonstrated in Nissl-stained sections of somatosensory cortex of mice and rats (e.g. Woolsey & Van der Loos 1970). More recently, brain sections cut parallel to the brain surface and processed for cytochrome oxidase or succinic dehydrogenase reveal that separate digits, and pads of the palm, are distinct as well (e.g. Dawson & Killackey 1987). In similar brain sections from primary somatosensory cortex of monkeys stained for myelin, each digit of the hand is apparent as a myelin-dense oval separated by narrow myelin-light septa (Jain et al 1998). In the brain of star-nosed moles, which have a nose with 22 fleshy appendages that are used in searches for food, cytochrome oxidase preparations reveal that there are three separate representations of the appendages or rays of the nose in cortex, with a cytochrome oxidase-dense band for each ray that is outlined by a narrow septum in each (Catania & Kaas 1995). Similar cytochrome oxidase bands for rays exist in the ventroposterior nucleus of the thalamus and in the trigeminal complex of the brainstem (Fig. 1). The general finding is that disruptions of the receptor sheet, caused by protuberances like the digits, separations in the skin, as between the lips, or folds, as between pads of the palm, are reflected in representations by narrow separating septa between extents of tissue representing continuous parts of the receptor sheet. The larger extents or cores of tissue are isomorphs of the body parts that collectively form representations. Cores receive the main sensory inputs, while septa receive callosal and other less direct connections (see Jain et al 1998).

Visual and auditory receptor surfaces do not have the splits and folds of the skin. Nevertheless, the retina does have the disruption of the nerve head or optic disc. In the layers of the lateral geniculate nucleus related to the contralateral eye, the nerve head is often clearly reflected by a rod-like, cell-poor discontinuity in the layers (Kaas et al 1973b). The optic disc is also apparent in sections of flattened primary visual cortex when inputs from one eye are labelled. An oval of cortex exists with inputs from the ipsilateral eye but not from the contralateral eye (e.g. Florence & Kaas 1992). This type of disruption occurs in area 17 even when the optic disc is in the monocular visual field, and thus the optic disc is represented in cortex with inputs only from the contralateral eye. In ground squirrels, with limited binocular overlap, the nerve head is an elongated horizontal strip in the monocular field of the nasal retina. The dense myelination of primary visual cortex, area 17 or V1, in these squirrels is disrupted by a narrow myelin-light septum marking the projection of the optic 'disc' (H. Rodman, personal communication 1999). Thus, neurons in cortex activated by retinal receptors above or below the nerve head are separated by a myelin-light septum in cortex.

Errors in the development of sensory surfaces
are transmitted to cortical maps

The number and arrangement of vibrissae on the muzzle of mice is typically constant across individuals, but occasionally differences appear. As already

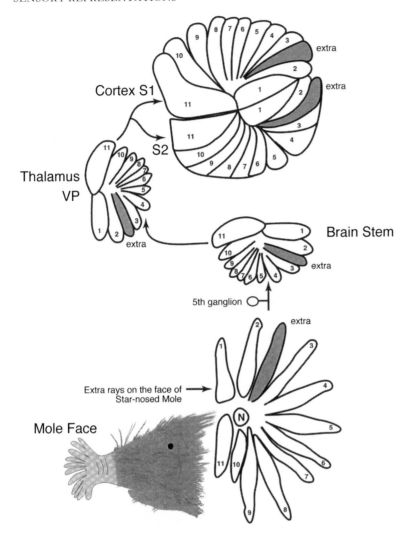

FIG. 1.   The reorganization of the somatosensory system of the star-nosed mole as the result of being born with an extra ray on the nose. Normally these moles have 22 appendages, 11 on each side of the face. Occasionally, one is born with one or two extra or less on each side. In star-nosed moles, the representation of the rays of nose are visible as dense band-like regions separated by narrow, light septa in cytochrome oxidase and other histological preparations. Any change in the number of rays is reflected in a change in the number of bands in the trigeminal nucleus of the brainstem, the ventroposterior (VP) nucleus of the thalamus, and in primary (S1) and secondary (S2) representations in somatosensory cortex. Because it seems unlikely that separate genetic codes for each brain location would suddenly emerge, or that a single genetic code would directly mediate changes in the nose and in at least four levels of the somatosensory system, we conclude that the change in the nose 'instructs' the development of the detailed somatototopy up the nervous system. See Catania & Kaas (1997b). N, nostril.

mentioned, the distribution of mystacial vibrissae on the face of mice is precisely matched by the distribution of morphologically distinct modules in cortex (Woolsey & Van der Loos 1970), termed barrels, with one barrel for each whisker. Van der Loos and co-workers noted that mice occasionally vary in number of whiskers, having one or two extra or less whiskers, and that such mice could be selectively bred to form strains with different patterns of vibrissae. Whatever the number of whiskers, that number was precisely reflected in the number of barrels in somatosensory cortex, and in the number of equivalent structures in the thalamus and brainstem (Van der Loos & Dörfl 1978, Welker & Van der Loos 1986). Thus, the spacing of receptors around whiskers in the skin determined how sensory maps were constructed.

Similar results were obtained when brain maps were related to natural variations in the number of mobile appendages that extend from the nose of the star-nosed mole. Normally, each side of the nose has 11 rays, and cortical and subcortical representations also have 11 bands, one for each ray (Catania & Kaas 1997a). However, one mole that was captured with 12 rays had 12 bands in these representations (Catania & Kaas 1997b). Subsequently, we found other moles with 12 rays or 10 rays, and they had the corresponding 12 or 10 bands in each representation (Fig. 1). Our interpretation of these results follows the early conclusion of Van der Loos & Dörfl (1978) that 'the cortical array is slaved to the peripheral array, as a result of a cascaded induction over three synaptic stations'.

In the visual system, a similar alteration of the receptor sheet occurs in Siamese cats where the contralaterally projecting retina extends 20° or so beyond normal. The abnormal extension of the contralaterally projecting retina results in a fusion of representations of the normal and the abnormal portions of the retina in the lateral geniculate nucleus (Guillery & Kaas 1971) and either an extension of the retinotopic map in area 17 to include the additional 20° of retina, or a suppression of this input (Hubel & Wiesel 1971, Kaas & Guillery 1973). The critical observation here is that when the contralaterally projecting retina is extended by the additional 20°, the representations in the lateral geniculate nucleus and sometimes cortex respond by representing the extra 20° of retina in a continuous retinotopic pattern (Fig. 2).

Modular subdivisions of sensory representations emerge when inputs differ in activity patterns

The best example of how cortical representations become subdivided by competing inputs with different activity patterns is the emergence of ocular dominance 'columns' or bands in area 17 of monkeys and cats. Inputs to layer 4 of area 17 divide up the sensory representation into nearly equal amounts of alternating bands activated by one eye or the other (e.g. Florence & Kaas 1992).

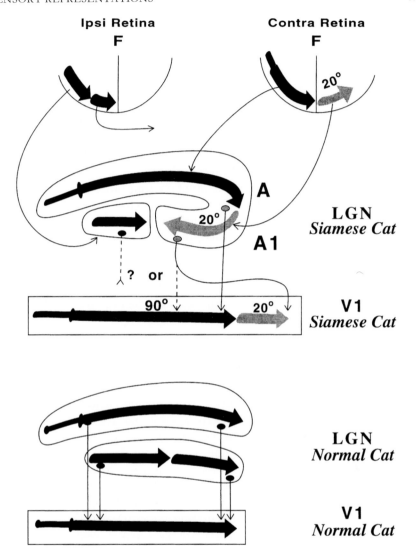

FIG. 2.   Retinotopically extended representations in the lateral geniculate nucleus (LGN) and primary visual cortex (V1) of Siamese cats. In Siamese cats, a narrow 20° strip of retina just temporal to normal visual fixation (F) wrongly projects to the contralateral rather than the ipsilateral LGN, thereby extending the continuous retinotopic pattern by 20°. Part of the A1 geniculate layer receives this abnormal input, and fuses with the normal A layer to form the extended representation. Because projections to V1 from the A1 layer no longer retinotopically match those from the A layer, the A1 projections develop to either become very sparse and weak, or to reorganize in cortex to form an extended representation. The potential for such reorganizations that partially correct a primary defect suggest a role for neural activity in forming central connections. See Kaas & Guillery (1973).

Such bands can be artificially induced in the optic tectum of frogs by adding an extra eye in development so that two eyes compete in the same structure (Constantine-Paton 1982). The band-like pattern in monkeys is a consequence of the nearly equal balance of inputs from the two eyes, and it changes to a dot and surround pattern in the part of area 17 representing peripheral vision where the contralateral eye inputs start to dominate. Computer simulations of the pattern suggest that a balance between factors grouping inputs from the same eye and factors superimposing retinotopic patterns from the two eyes produce the cortical patterns (Tanaka 1991; also see Constantine-Paton 1982). The dot pattern of cytochrome oxidase blobs with 'K' channel inputs in area 17 of monkeys (see Casagrande & Kaas 1994) could be a result of the same process. Inputs to area 18 or V2 of monkeys are similarly segregated by type into sets of three classes of alternating bands (see Livingstone & Hubel 1988). The middle temporal visual area, MT, also has a tilework of modules. In area 3b of somatosensory cortex of monkeys, we see a segregation of alternating bands of neurons activated by either slowly adapting or rapidly adapting cutaneous afferents (Sur et al 1984). Finally, the formation of sublayers, such as those for either 'on' or 'off' centre ganglion cells of the retina in area 17 of tree shrews, resembles the process of forming modules.

Parts of the receptor array can capture more than their share of cortical space

In general, the proportions of sensory representations reflect the innervation densities of different sectors of the receptor sheet (e.g. see Welker & Van der Loos 1986). An obvious exception is the disproportionately large representation of emitted sonar frequencies in auditory cortex of echolocating bats. More recently, it has also become clear that the cortical representation of the fovea in monkeys is larger than would be predicted from ganglion cell densities (Azzopardi & Cowey 1993). Finally, the receptors of ray 11 of the star-nose mole are over-represented in cortex in a manner that suggests a tactile 'fovea' (Catania & Kaas 1997a). Each of the 11 rays of each side of the nose is covered with specialized sensory organs. Yet, ray 11, which hangs over the mouth, is behaviourally most important in food identification, and this has more than twice the expected cortical territory.

Adjoining sensory representations have congruent borders

Cortical representations tend to be topographically matched along their common borders, i.e. they have congruent borders (Allman & Kaas 1975). In two adjacent visual representations with congruent borders, V1 and V2 for example, adjacent locations on each side of the border will represent similar parts of the visual field. Likewise, in the adjoining somatosensory representations in areas 3b and 1 of somatosensory cortex of monkeys, neurons just across the common border in hand cortex are activated by comparable parts of the palm (e.g. Kaas et al 1979). In contrast, the border between area 1 and area 2 for the hand is matched along the fingertips. Auditory areas may join and match along representations of high or low frequencies, or along the frequency range (Kaas & Hackett 1998). The situation becomes more complex where smaller representations border only parts of larger representations and border several other representations as well, and it becomes difficult to fully maintain congruent borders. Yet, even under these circumstances, borders remain remarkably congruent. In monkeys, for example, two smaller somatosensory cortical areas adjoin the lateral part of area 3b representing the face, and they adjoin each other. Each smaller representation, S2 and the parietal ventral area (PV), has a different short congruent border along different parts of the face in area 3b, and a congruent common border along the face and hand representations (e.g. Krubitzer & Kaas 1990). S2 also borders the ventral somatosensory area, VS, along mirror representations of the hand and foot (Cusick et al 1989). Similar matches occur in visual cortex of monkeys where small visual areas, such as the dorsomedial (Allman & Kaas 1975) and medial (Allman & Kaas 1976) visual areas congruently adjoin parts of V2 representing paracentral parts of the lower visual quadrant and congruently join each other along representations of more peripheral parts of the lower visual quadrant. Congruent borders are common and they are maintained or approximated even where it is difficult to do so with seemingly incompatible neighbours. This is done by the presence of rapid representational transitions across part of the receptor sheet. While boundaries are more easily approximated for smaller representations with larger receptive fields, the common observation is that areas have representationally matched borders.

Developmental and functional implications of common features of sensory representations

Much of what we see in sensory representations could developmentally emerge as the result of the interplay of a few factors, especially the action of some sort of chemical guidance and attraction (Sperry 1963), so that an overall but probably a crude pattern of connections between structures is established, followed by a refining role of neural activity patterns to create detailed order, disruptions, disproportions, modules and congruent borders (see Kaas 1988, 1993). Thus, the septa that emerge between the representation of the fingers in somatosensory cortex of monkeys are seen as no-man zones formed between populations of neurons with slightly different activity patterns starting even before birth as skin afferents first become functional. The formation of extra barrels in S1 of mice with extra whiskers or bands in S1 of moles with extra nose rays seems incompatible with any explanation that does not involve the transfer of information from the face to the brain, especially when one realizes that the brain changes occur across at least two levels of subcortical processing, and at least at one level of cortical processing. Genetic instructions for matched changes in the face and three to five places in the brain would not occur independently in one generation. Instead, information must come from the receptor sheet. The most probable source of information is in the neural activity patterns (see Shatz 1990), although there are arguments against this premise. Since neurons also transport substances, this is another potential source of information that could guide the matching of systems. If borders between representations are at least originally somewhat imprecise, coincident activations of neurons within a border zone would tend to align representations and match borders, especially if one representation develops slightly later than the other.

The matching of representations across areas requires that information from one area spills across the border to the next. Sensory representations that develop with instructions from the receptor sheet would require reduced genetic instructions, and would be quite adaptable and responsive to the environment. Such representations would also be highly functional (Kaas 1997). In topographic maps, neurons that need to interact are generally close together, and thus the lengths of interconnections would be reduced. Topographic maps permit the formation of effective local circuits. While separate representations do interconnect, relatively few such longer connections are needed, and the congruence of representations along common borders shortens these connections, especially for representations that are extensively aligned such as V1 and V2. Neurons that are seldom coactivated and seldom interact are connectionally isolated by septa. The formation of modules for different classes of inputs further groups interacting neurons. Inputs with earlier or more activity acquire more cortical space, and more 'afferent magnification'. Overall, a system of topographic maps that is highly responsive to activity patterns during development is well designed.

On the other hand, if sensory representations are created by the interplay of only a few interacting factors, they are unlikely to be optimally designed. Maps may contain features of little or no functional significance; some features may simply be necessary outcomes when selection favours other features. In other words, many features of sensory maps are probably developmentally linked, so that they cannot be independently regulated. Thus, all aspects of maps may not have functional significance. Rather than try to explain all map features in terms of function, we might consider that the balance of developmental influences leads to a generally useful representation.

Acknowledgements

Research by the author is funded by the Institute of Neurological Disorders grant no. 16446 and National Eye Institute grant no. EY02686.

References

Allman JM, Kaas JH 1975 The dorsomedial cortical visual area: a third tier area in the occipital lobe of the owl monkey (*Aotus trivirgatus*). Brain Res 100:473–487

Allman JM, Kaas JH 1976 Representation of the visual field on the medial wall of occipital-parietal cortex in the owl monkey. Science 191:575–575

Azzopardi P, Cowey A 1993 Preferential representation of the fovea in primary visual cortex. Nature 361:719–721

Casagrande VA, Kaas JH 1994 The afferent, intrinsic, and efferent connections of primary visual cortex in primates. In: Peters A, Rockland KS (eds) Cerebral cortex, vol 10: Primary visual cortex in primates. Plenum Press, New York, p 201–259

Catania KC, Kaas JH 1995 The organization of the somatosensory cortex of the star-nosed mole. J Comp Neurol 351:549–567

Catania KC, Kaas JH 1997a Somatosensory fovea in the star-nosed mole: behavioral use of the star in relation to innervation patterns and cortical representations. J Comp Neurol 387:215–233

Catania KC, Kaas JH 1997b The mole nose instructs the brain. Somatosens Mot Res 14:56–58

Catania KC, Lyon DC, Mock, OB, Kaas JH 1999 Cortical organization in shrews: evidence from five species. J Comp Neurol 410:55–72

Constantine-Paton M 1982 The retinotectal hookup: the process of neural mapping. In: Subtelny S, Green PB (ed) Developmental order: its origin and regulation. Alan R Liss, New York, p 317–349

Cusick CG, Wall JT, Felleman DJ, Kaas JH 1989 Somatotopic organization of the lateral sulcus of owl monkeys: area 3b, S-II, and a ventral somatosensory area. J Comp Neurol 282:169–190

Dawson DR, Killackey HP 1987 The organization and mutability of the forepaw and hind paw representations in the somatosensory cortex of neonatal rat. J Comp Neurol 256:246–256

Florence SL, Kaas JH 1992 Ocular dominance columns in area 17 of Old World macaque and talapoin monkeys: complete reconstructions and quantitative analyses. Vis Neurosci 8:449–462

Guillery RW, Kaas JH 1971 A study of normal and congenitally abnormal retinogeniculate terminations in cats. J Comp Neurol 143:73–100

Hubel DH, Wiesel TN 1971 Aberrant visual projections in the Siamese cat. J Physiol (Lond) 218:33–62

Jain N, Catania KC, Kaas JH 1998 A histologically visible representation of the fingers and palm in primate area 3b and its immutability following long-term deafferentations. Cereb Cortex 8:227–236

Johnson JI 1990 Comparative development of somatic sensory cortex. In: Jones EG, Peters A (eds) Cerebral cortex vol 8a: Comparative structure and evolution of cerebral cortex, part 2. Plenum Press, New York, p 335–449

Kaas JH 1988 Development of cortical sensory maps. In: Rakic P, Singer W (eds) Neurobiology of neocortex. Wiley, New York, p 115–136

Kaas JH 1993 The evolution of multiple areas and modules within neocortex. Perspect Dev Neurobiol 1:101–107

Kaas JH 1997 Topographic maps are fundamental to sensory processing. Brain Res Bull 44:107–112

Kaas JH, Guillery RW 1973 The transfer of abnormal visual field representations from the dorsal lateral geniculate nucleus to the visual cortex in Siamese cats. Brain Res 59:61–95

Kaas JH, Hackett TA 1998 Subdivisions of auditory cortex and levels of processing in primates. Audiol Neurootol 3:73–85

Kaas JH, Harting JK, Guillery RW 1973a Representation of the complete retina in the contralateral superior colliculus of some mammals. Brain Res 65:343–346

Kaas JH, Guillery RW, Allman JM 1973b Discontinuities in the dorsal lateral geniculate nucleus corresponding to the optic disc: a comparative study. J Comp Neurol 147:163–179

Kaas JH, Nelson RJ, Sur M, Lin CS, Merzenich, MM 1979 Multiple representations of the body within 'S-I' of primates. Science 204:521–523

Krubitzer LA, Kaas JH 1990 The organization and connections of somatosensory cortex in marmosets. J Neurosci 10:952–974

Livingstone M, Hubel D 1988 Segregation of form, color, movement, and depth: anatomy, physiology, and perception. Science 240:740–749

Merzenich MM, Kaas JH 1980 Principles of organization of sensory-perceptual systems in mammals. In: Sprague JM, Epstein AN (eds) Progress in psychobiology and physiological psychology. Academic Press, New York, p 1–42

Shatz CJ 1990 Impulse activity and the patterning of connections during CNS development. Neuron 5:745–756

Sperry R 1963 Chemoaffinity in the orderly growth of nerve fiber patterns and connections. Proc Natl Acad Sci USA 50:703–710

Sur M, Wall JT, Kaas JH 1984 Modular distributions of neurons with slowly adapting and rapidly adapting responses in area 3b of somatosensory cortex in monkeys. J Neurophysiol 51:724–744

Tanaka S 1991 Theory of ocular dominance column formation. Biol Cybern 64:263–272

Van der Loos H, Dörfl J 1978 Does the skin tell the somatosensory cortex how to construct a map of the periphery? Neurosci Lett 7:23–30

Van Essen DC, Drury HA 1997 Structural and functional analyses of human cerebral cortex using a surface-based atlas. J Neurosci 17:7079–7102

Welker E, Van der Loos H 1986 Quantitative correlation between barrel-field size and the sensory innervation of the whisker pad: a comparative study in six strains of mice bred for different patterns of mystacial vibrissae. J Neurosci 6:3355–3373

Woolsey TA, Van der Loos H 1970 The structural organization of layer IV in the somatosensory region (SI) of mouse cerebral cortex. The description of a cortical field composed of discrete cytoarchitectonic units. Brain Res 17:205–242

DISCUSSION

Purves: The idea that peripheral receptor sheets determine the organization of the relevant parts of the central nervous system is widespread, but I'm not sure it's correct. For example, consider a person born with an extra digit. Is it reasonable to suggest that the ability to process all the relevant information in the cortex that pertains to the extra digit, is generated from the periphery?

Kaas: What I'm suggesting is that the complete sensorimotor system is flexible in development and adjusts to changes in sensory input without any change in genetic instruction. In the case where the number of whiskers on the face of the mouse varies, for example, even the number of whiskers apparently is not genetically specified, but genes that affect the folding of the face during development influence the number of whiskers. Thus, the number of whiskers is indirectly controlled, and this number alters the developing patterns of motor and sensory innervation, and this alters the rest of the sensorimotor system. Matching changes throughout the system that emerge from one generation to the next cannot be independently specified.

Purves: But surely it is asking a lot for retrograde information from the extra digit as such to control all this stuff.

Kaas: It is asking a lot, but control by the receptor sheet seems to be the most reasonable possibility. Neurons are designed to transfer information from one level to the next. They do this with activity patterns and by transporting molecules. It is possible that a chemical signal in the skin is transported over several stages in the somatosensory system, and that this signal organizes structures. The more likely possibility is that neural activity does this. The alternative of local, independent control at each level seems improbable.

Purves: It depends how far back you go in development. Bill Harris originally threw doubt on this hypothesis when he transplanted eyes from newts that were tetrodotoxin-sensitive to newts that were tetrodotoxin-resistant (Harris 1980, 1981, 1984). The results showed that an inactive eye develops perfectly good retinotectal connections. More recently, Larry Katz and his collaborators (personal communication, 1999) have shown that enucleated ferrets develop ocular dominance columns. There's some reason, therefore, to be suspicious about the role of neural activity in cortical organization, at least as that role is usually conceived.

Kaas: I agree. There is a lot of evidence that argues against activity. The same situation occurs in the somatosensory system, i.e. activity doesn't seem to have a role in the barrel field formation. But one can find ways to question these sorts of experiments, so one is left considering how to explain these results. We need an alternative that explains these multiple levels of change.

O'Leary: If the event is genetically controlled, I suspect that one would need to provide a somewhat elaborate scheme to explain the coordinated changes in gene expression at multiple levels of the neuroaxis. One observation that is relevant to this issue is the demonstration that cadherin 6, 8 and 11 are expressed in specific sensory systems at different levels of the neuroaxis, in each of the relevant thalamic, midbrain and hindbrain nuclei, as well as in the cortical fields, at the appropriate times in development (Suzuki et al 1997). The mechanisms that control such expression of a specific molecule in various components of a given sensory system at multiple levels of the neuroaxis, which otherwise have no other relationships to one another, at least in terms of regulatory gene expression or developmental origins, are either complex, or coincident. I would be interested to hear speculation on a potential mechanism that might control this type of coordinated gene expression.

Puelles: One can hypothesize that morphogenetic fields in different regions of the brain are involved, as well as several general rules of cell behaviour, which may be independently implemented at each locus by different sets of cells, potentially leading to multiple coherent interpretations of information available in the diverse fields. Emergent aspects may turn up variously in such a scenario, i.e. from changing the rules, adding or subtracting elements that implement the rules, or even changing the number, size or other relevant parameters of the fields. Segmentation, dorsoventral patterning and other regionalization processes of the neural tube wall apparently establish the number, size and positions of the relevant primary or secondary fields. Excess production of neurons and size matching by means of trophic factor requirements is one potential general rule affecting all fields independently, although the resultant equilibria across neural pathways will tend to be interdependent. Number and place of collaterals produced by axons within a pathway may imply several general navigational rules and sets of decision points operating both across diverse fields and/or within any field. Gradients of attractor or repellent signalling substances imply complex rules acting possibly independently within separate fields, in ways which may allow subsequent simple or complex matching between fields. Activity-related reorganization of synaptic contacts imply other sorts of rules, etc. Somehow, the net result of these multiple mechanisms is that phenomena in one field are connected in a patterned, functionally efficient way to phenomena occurring in other fields. Such effects may have selective value in evolution. Many of the relevant processes may be redundant and serve ordinarily as buffering mechanisms for developmental error, but essentially we have multistable systems, which may change abruptly from one equilibrium state to another, due to subtle modulation of genetic or epigenetic influences.

Levitt: This doesn't explain the matching at the molecular level, because the matching at the molecular level occurs earlier than the wave of interactions that occur in each of the projection stations. One answer is that each gene has multiple enhancers or promoters, and so different combinations may regulate expression in each of those systems.

O'Leary: But they would have to have a common signal.

Goffinet: Do the differences you are alluding to become apparent prior to any input or connection?

O'Leary: Yes. John Rubenstein presented evidence yesterday that several genes, for example cadherin 6, exhibit their normal differential expression patterns in the cortex of *Gbx2* mutant mice that lack thalamocortical input. Therefore, at least for the genes that he described, the mechanism controlling their differential expression patterns is intrinsic to cortex.

Levitt: Takeichi's laboratory showed that as early as embryonic day (E) 14.5, the distribution patterns of cadherin 6 and 11 are apparent before connections are formed, so there is independent regulation at this stage (Redies & Takeichi 1996).

Goffinet: But there are some connections that are formed early, and these may play a role.

Levitt: Certainly not those from the thalamus to the cortex.

Rakic: There is no question that input also plays a role, particularly during the second stage when cells are already in their proper positions and have received proper input. Thus, the pattern of connectivity depends on the expression of genes that serve as markers which attract proper input. However, the number of ipsilateral and contralateral connections that serve the same function is determined by competition (Meissirel et al 1997).

Karten: I would like to throw in a comment in relation to Jon Kaas' presentation and Dale Purves' response. We are mostly neurobiologists at this meeting, and we tend to view evolution and the structure of the brain as if the brain is driving everything. One of the most important points that emerges from Jon's talk, is that the brain is just another organ. The issue is that the brain doesn't evolve by itself, and all the regulatory events that occur within the brain are not driven primarily by the brain. Novelties within the brain may therefore not only reflect novelties that initiate in the brain.

Kaas: What I was trying to point out is that we can see effects in the brain that can be best explained by changes in neural activity patterns. If we look at visual cortex in Siamese cats, the representation can be extended by as much as $20°$ by a realignment of inputs relayed from abnormal projections from the retina. Cortex adjusts by developing a retinotopic map by altering the normal pattern of geniculocortical connections. Individual cats vary so that an alternative correction is to suppress geniculocortical inputs that terminate in 'normal' locations but would disrupt retinotopic organization. Both adjustments reflect the normal potential for developmental plasticity that is already built into the system. The adjustments are unrelated to the genetic change in coat and eye pigment that produces Siamese cats with misdirected retinal projections.

Rakic: Within a given system or modality, basic connectivity is established first. Only then can interactions take place, e.g. activity-dependent selective elimination of neurons, axons and synapses. In this respect, connections are sharpened and diversity is produced by these cell interactions. We would avoid misunderstandings if we defined what we meant by cytotectonic areas or fields. Some of the cortical regions are not in the same categories, and some of these categories depend on interactions between cells. For example, although the somatosensory region is defined early on, one could imagine how barrel fields could be changed because their final pattern depends on connectivities.

Molnár: But Jon Kaas was saying that if you modify the crossing of the retinal projection in the optic chiasm and modify the input, you generate two different outcomes in the cortical representation, i.e. the Boston and Midwestern patterns. However, we still don't know why the brain generates these two outcomes. Why is it that the map is suppressed in one type and reversed in the other? These changes are associated with changes of the thalamocortical projections. How do thalamocortical projections change as the result of the altered flow of memory input? Perhaps it's too early to bury the activity story, because there is evidence which suggests that as soon as thalamic fibres arrive in the cortex they can elicit activation. If you look at how they enter the cortex in rodents just after birth, you see that they are regular (Agmon et al 1995); but if you look a few days before when they line up within the intermediate zone, you see transient side branches (Naegele et al 1988) and they transmit activation patterns (Higashi et al 1996) that could self-organize their entry into the cortex. Perhaps this is one of the mechanisms by which cortical representations can reverse, expand or get suppressed. There is an example for such reversal. For instance, between the representation of the lateral geniculate nucleus (LGN) and of the primary visual cortex polarity changes. Connolly & Van Essen (1984) proposed that the reversal of the representation in the cortex can only be explained by the decunation of the optic radiation fibres at some point in the pathway in a mediolateral, but not an anteroposterior dimension. Nelson & LeVay (1985) applied two different tracers into the cat LGN in pairs, anteroposterior and mediolateral, and found that the mediolateral labelling revealed a twist in the pathway about 200–$500\,\mu m$ below the cortex, which could have been in the subplate during development. Perhaps the anatomical substrate of this rearrangement was in the side branches of the thalamocortical projections (Naegele et al 1988) during the period when they start to transmit these early activation patterns.

Pettigrew: I'm also keen on activity as an epigenetic mechanism. We can't bury it just because Bill Harris used a blocker of the sodium channel and failed to find developmental effects.

Purves: I don't want to bury it, just to urge a certain amount of caution in considering its developmental role.

Pettigrew: It is possible to get both Boston and Midwestern patterns in the same Siamese cat, depending on vertical eccentricity (Cooper & Blasdel 1980). I don't wish to go into details, but there is a lot of explanatory power in patterns of activity. There is a experiment that has been performed on marmoset monkeys that is difficult to explain genetically, but is explicable epigenetically. When the marmoset monkey is born, it has beautiful ocular dominance columns. The explanation for this is that there are two independent sources of activity, i.e. the two eyes, and the brain can work out that the calcium waves passing across the two eyes are separated in time and space. If you let the marmoset see the world, those ocular dominance columns disappear. This is paradoxical. The marmoset monkey is unusual. It's a dwarf, and its eyes are close together. It has inherited a system from larger monkeys, and my interpretation is that when it starts seeing the world, the beautiful separation of the two eyes that it had *in utero* has now disappeared. The disparities are so small in the real world that unless you present stimuli from 0.5 cm and infinity, there is not enough asynchrony in the cortex. So, I can explain this epigenetically, but not genetically. At the moment, we have to accept that the brain can generate complex patterns of activity, some of which may involve sodium channels or calcium channels. Those emergent properties of the way these processes interact may overlap with morphogenetic fields.

Purves: The point I'm making is simply that this is a open question. The central effects of having an extra whisker don't really argue against the idea of simultaneous modifications at several levels of the system.

Pettigrew: Activity can turn genes on and off. I'm not saying that genes aren't involved.

Purves: I would like to ask Jon Kaas a question on a different topic. Your topographic maps bring to mind the plasticity experiments of Merzenich et al (1983a,b, 1987, 1990). Now that you have shown an anatomical correlate of digits in monkey S1, would you expect this map to change in such experiments?

Kaas: We have investigated this possibility. In the barrel field of rats, each barrel is best activated by one whisker, but it can be less strongly activated by other whiskers, so a barrel is not exclusive for one whisker. In a similar manner, each of the cytochrome oxidase territories in the cortex of monkeys has neurons best activated or only activated by stimulating a specific digit. However, the sources of activation can be changed, especially in sensory deprivation experiments. When this is done so that neurons become activated by other digits or even the face, the cytochrome oxidase territories remain unchanged. I assume that after a certain developmental phase, a hard-wired framework is in place that is reflected in the cytochrome oxidase, and myelin, expression. This suggests that even when cortical somatotopy is greatly altered, most of the connectional framework remains in place.

Purves: These results are important in clearing up the discrepancy between the relative lack of plasticity after the end of critical periods in the visual system, and the apparently greater degree of ongoing plasticity that has been demonstrated in the somatosensory system.

Rubenstein: The discussion on the mapping of the sensory fields is not complete without a little discussion of the olfactory maps. It appears that activity-dependent processes do not have a major role at the early stages of mapping the primary olfactory axons onto a two-dimensional spatial map of the olfactory bulb. There clearly is a molecular mechanism to do this.

Levitt: What's the nature of the spatial information that's generated from an odour?

Rubenstein: The spatial arrangement of olfactory information on the surface of the olfactory bulb generates a topographic order to the system, allowing the animal to reproducibly identify which part of its brain is being stimulated.

Levitt: What you want to create is almost an immune system in the olfactory system. You want to give the organism a chance to respond to almost any odour it might come across, but surely the animal doesn't care whether it's coming from a particular point in space, just general direction.

Rubenstein: All I'm saying is that, as I understand, one can create a two-dimensional representation of smell on the olfactory bulb in a way that's reproducible from animal to animal. This may be important in terms of connectivity, because one would want a smell associated with a dangerous animal, for instance, to go to a part of the amygdala. I have no idea whether such molecular-driven maps are required for other sensory modalities.

O'Leary: Just to paraphrase what John Rubenstein has said, in the olfactory system finely ordered connections appear to be generated by just using purely molecular information. This is in a system where spatial information is not necessary in the sensory stimulus. In contrast, in the visual system spatial information is required in the sensory stimulus, but even so, a substantial degree of topography is generated in the absence of activity (for review of both systems, see O'Leary et al 1999).

Levitt: Let's take the example of the experiments of Patricia Gaspar and colleagues, in which they demonstrated a dissociation in the patterning from the peripheral receptors to the patterning that occurs in the barrel cortex, where presumably there is altered activity due to the changes in serotonin levels (Vitalis et al 1998, Cases et al 1996). This dissociation can be restored during a critical period of time, even though the cortical maps have been altered.

O'Leary: In the visual system, activity blockade in the colliculus or in the retina during the period when the retinocollicular map forms, does not prevent the development of a fairly well-ordered map (O'Leary et al 1999).

Kaas: There's no doubt that an order is created, but an order doesn't depend on activity, and there is no evidence that the order is perfect.

Boncinelli: Generally speaking, earlier stages of development reveal more genetically determined and relatively fixed biological schemes, whereas progressively later stages reveal more and more regulated, epigenetic and experience-driven schemes.

Krubitzer: If you are interested in cortical specification within the lifespan of an individual, or in a species over time, you can consider phenotypic variation of cortical fields across mammals, or you can consider pure evolution. If you are interested in the latter, then you have to examine genetic change. If you are interested in the former, you have to invoke both genetic and epigenetic events to explain variation, because a genome doesn't develop in isolation.

References

Agmon A, Yang LT, Jones EG, O'Dowd DK 1995 Topological precision in the thalamic projection to neonatal mouse barrel cortex. J Neurosci 15:549–561

Cases O, Vitalis T, Seif I, De Maeyer E, Sotelo C, Gaspar P 1996 Lack of barrels in the somatosensory cortex of monoamine oxidase A-deficient mice: role of a serotonin excess during the critical period. Neuron 16:297–307

Connolly M, Van Essen D 1984 The representation of the visual field in parvicellular and magnocellular layers of the lateral geniculate nucleus in the macaque monkey. J Comp Neurol 226:544–564

Cooper ML, Blasdel GG 1980 Regional variation in the representation of the visual field in the visual cortex of the Siamese cat. J Comp Neurol 193:237–253

Harris WA 1980 The effects of eliminating impulse activity on the development of the retinotectal projection in salamanders. J Comp Neurol 194:303–317

Harris WA 1981 Neural activity and development. Annu Rev Physiol 43:689–710

Harris WA 1984 Axonal pathfinding in the absence of normal pathways and impulse activity. J Neurosci 4:1153–1162

Higashi S, Molnár Z, Kurotani T, Inokawa H, Toyama K 1996 Functional thalamocortical connections develop during embryonic period in the rat: an optical recording study. Soc Neurosci Abstr 22:1976

Meissirel C, Wikler KC, Chalupa LM, Rakic P 1997 Early divergence of magnocellular and parvocellular functional subsystems in the embryonic primate visual system. Proc Natl Acad Sci USA 94:5900–5905

Merzenich MM, Kaas JH, Wall JT, Nelson RJ, Sur M, Felleman DJ 1983a Topographic reorganization of somatosensory cortical areas 3b and 1 in adult monkeys following restricted deafferentation. Neuroscience 8:33–55

Merzenich MM, Kaas JH, Wall JT, Sur M, Nelson RJ, Felleman DJ 1983b Progression of change following median nerve section in the cortical representation of the hand in areas 3b and 1 in adult owl and squirrel monkeys. Neuroscience 10:639–665

Merzenich MM, Nelson RJ, Kaas JH et al 1987 Variability in hand surface representations in areas 3b and 1 in adult owl and squirrel monkeys. J Comp Neurol 258:281–296

Merzenich MM, Recanzone GH, Jenkins WM, Nudo RJ 1990 How the brain functionally wires itself. In: Arbib MA, Robinson JA (eds) Natural and artificial parallel computation. MIT Press, Cambridge, MA, p 177–210

Naegele JR, Jhaveri S, Schneider GE 1988 Sharpening of topographical projections and maturation of geniculocortical axon arbors in the hamster. J Comp Neurol 277:593–607

Nelson SB, LeVay S 1985 Topographic organization of the optic radiation of the cat. J Comp Neurol 240:322–330

O'Leary DDM, Yates PA, McLaughlin T 1999 Molecular development of sensory maps: representing sights and smells in the brain. Cell 96:255–269

Redies C, Takeichi M 1996 Cadherins in the developing central nervous system: an adhesive code for segmental and functional subdivisions. Dev Biol 180:413–423

Suzuki SC, Inoue T, Kimura Y, Tanaka T, Takeichi M 1997 Neuronal circuits are subdivided by differential expression of type-II classic cadherins in postnatal mouse brains. Mol Cell Neurosci 9:433–447

Vitalis T, Cases O, Callebert J et al 1998 Effects of monoamine oxidase A inhibition on barrel formation in the mouse somatosensory cortex: determination of a sensitive developmental period. J Comp Neurol 393:169–184

How does evolution build a complex brain?

Leah A. Krubitzer

Center for Neuroscience and Department of Psychology, 1544 Newton Court, University of California, Davis, CA 95616, USA

Abstract. To understand how complex brains evolve one can examine a variety of the products of the evolutionary process and then infer the mechanisms that generate the differences observed. We address this issue using a number of techniques. We combine neurophysiological recording techniques with neuroanatomical tracing techniques and histochemical methods in an effort to accurately determine the functional subdivisions of the neocortex in a variety of mammals. By using these techniques we can determine common features of neocortical organization, or common cortical areas, which are considered homologous. We can observe modifications to patterns of cortical organization, or to cortical fields specifically, that are independently evolved and generally related to morphological and behavioural specializations. Comparative studies have led us to consider the development of the neocortex and the specific changes in developmental mechanisms that might account for the observed changes in extant adults. Both comparative studies and developmental studies allow us to formulate hypotheses regarding how the neocortex is constructed in the life of an individual, and in a lineage over time.

2000 Evolutionary developmental biology of the cerebral cortex. Wiley, Chichester (Novartis Foundation Symposium 228) p 206–226

Comparative work on a number of different mammals indicates that changes in the size of the cerebral cortex and in the number of its functional subdivisions are perhaps the most dramatic alterations to the mammalian brain in evolution (Stephan et al 1988, Krubitzer 1995). Indeed, there is over a 3000-fold difference in the size of the brain of the smallest mammals, some shrews and mice, and that of some cetaceans, such as dolphins and whales (see Manger et al 1998 for review; Fig. 1). Although the precise relationships between structure, function and behaviour are often difficult to understand, an increase in cortical surface area and number of interconnected cortical fields is generally associated with an increase in sensory, perceptual, cognitive and behavioural complexity. An obvious question is: how are more cortical fields added in evolution?

Our laboratory has addressed this question by comparing the brains of a variety of mammals that represent major branches of evolution (Fig. 2) using a number of techniques to subdivide the neocortex. These techniques include multiunit

mouse   **dolphin**

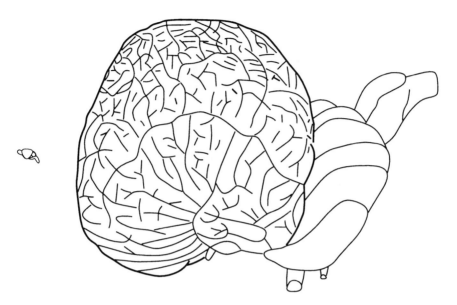

FIG. 1.   A mouse brain and dolphin brain drawn to scale. There is not only a large difference in the size of the neocortex, but also in the number of functional subdivisions that reside therein. In this and the following figures, rostral is to the left and dorsal is to the top.

electrophysiological recordings that allow us to sample a large extent of the neocortex, and to assign different modalities to different regions of the cortex. In conjunction with this, the architecture of the cortex is examined in the same animals and physiological boundaries are correlated with architectonic distinctions. Finally, the corticocortical, interhemispheric and subcortical connections of individual fields are examined to determine the unique pattern of interconnections that each field possesses.

While we cannot study evolution directly, by examining the products of the evolutionary process we can ascertain which aspects of cortical organization are common to all mammals, which features are unique to certain species and what types of modifications to the brain are made. In this way, we can make inferences about the evolutionary process, and the constraints placed on evolving brains. Because the evolution of the neocortex is actually the evolution of developmental programmes that generate the adult form, upon which selection operates, a second approach to understanding cortical field evolution is to study the developmental mechanisms that contribute to area specification (Killackey 1990).

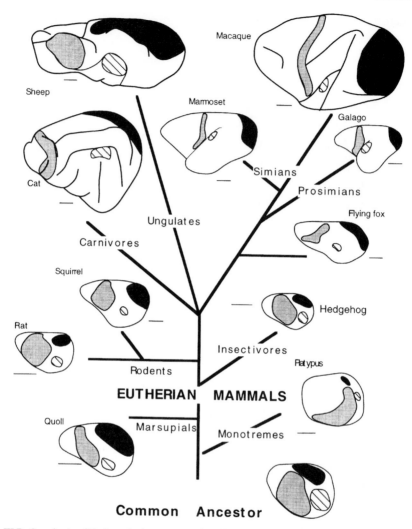

FIG. 2. A simplified evolutionary tree that depicts the major orders of mammals and the representative brain organization of common fields found in some of the species within each order. By comparing the brain organization across species and determining the common patterns of cortical field organization, it is possible to infer the organization of the common ancestor of all mammals. Once this is established, hypotheses regarding the changes in cortical organization that have occurred across lineages is possible. The black region denotes the primary visual cortex (V1), the striped region denotes the primary auditory cortex (A1) and the shaded region denotes the primary somatosensory cortex (S1). Note that the relative locations of these fields with respect to each other have shifted. Also, the relative amount of cortex assumed by these fields is often different for different animals. Finally, in animals with a greatly expanded neocortex (e.g. macaque monkeys), S1, A1 and V1 have moved far apart and new fields have been interspersed between these old fields.

This chapter outlines the cortical organization in a variety of different species, illustrates the features of organization that have been retained from the common ancestor and discusses the modifications to cortical organization that have occurred in different lineages. These observations allow us to evaluate the viability of current hypotheses regarding the development of the neocortex, and to propose the types of changes that might have occurred in developing brains during evolution to account for species differences.

Monotremes and marsupials

Electrophysiological recordings, coupled with architectonic analysis and studies of connections in two of the three species of extant monotremes, demonstrate that these animals have a constellation of cortical fields that represent the visual, auditory, somatosensory and, in the case of the platypus, the electrosensory epithelium (Krubitzer et al 1995). In both species, a primary somatosensory area, S1, can be readily identified with the foot represented most medially, followed by representations of the hindlimb, lower trunk, upper trunk, forelimb and face in a mediolateral progression (Fig. 3). S1 is coextensive with a myelin dark and cytochrome oxidase-dense field that receives inputs from the ventral posterior nucleus of the thalamus. Two additional representations of the somatosensory epithelium have also been observed and are termed the rostral field (R) and the parietal ventral area (PV). R contains a complete representation of deep receptors, while PV contains a complete representation of cutaneous receptors. Both species also contain at least one visual area, hypothesized to be the primary visual area (V1), and one auditory area, possibly A1 (Krubitzer 1998).

The most striking aspects of cortical organization in monotremes are the geographic arrangement of cortical fields and the cortical magnification of highly specialized peripheral body parts. For instance, V1 is just medial to the foot representation in S1, and auditory cortex is almost completely embedded in somatosensory cortex. In the platypus, the size of the bill representation in the neocortex is enormous. Indeed, across all of the representations examined, the representation of the bill assumes approximately 75% of the entire neocortex (Fig. 3).

Marsupials represent another early branch of evolution, since their ancestors diverged either with or slightly after the ancestors of extant monotremes (Westerman & Edwards 1992, Kirsch & Mayer 1998; Fig. 2). Our laboratory (Huffman et al 1999), as well as other laboratories (e.g. Sousa et al 1978, Beck et al 1996, see Johnson 1990, Rowe 1990 for review), demonstrate that like extant monotremes, marsupials possess a constellation of cortical areas that includes S1, S2/PV, R, V1, A1 and M (motor cortex). These fields have similar architecture, as well as similar patterns of connections from other cortical fields and from the

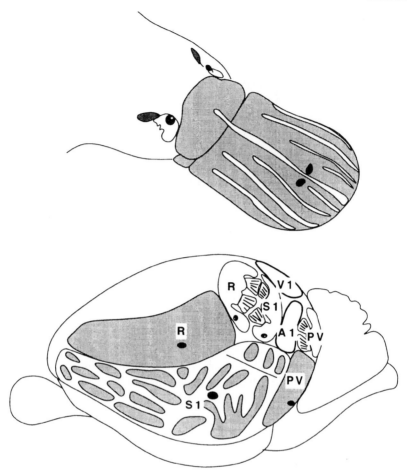

FIG. 3. The bill of the platypus (top) is highly specialized and contains both mechanosensory receptors (shaded) and electrosensory receptors (white) that are aligned in rostrocaudal stripes on the bill. The representations of the bill in the neocortex in the three somatosensory fields identified (bottom) is large and assumes most of the entire extent of the cortical sheet. In the primary somatosensory cortex (S1), representations of mechanosensory (shaded) and mechanosensory + electrosensory (white) are interdigitated with one another. In the rostral field (R), only mechanosensory and deep receptors are represented (shaded). In the parietal ventral area (PV) only mechanosensory receptors are represented (shaded). The other body part representations occupy far less of the neocortex. Primary visual cortex (V1) and auditory cortex (A1) are in locations that are unusual compared to other mammals.

thalamus (see Krubitzer 1995 for review). Like monotremes, the differences in cortical organization are, to a large extent, a reflection of morphological specialization. For instance, the dunnart is a small carnivorous marsupial that has a well-developed visual system which dominates the central nervous system. In addition to large eyes, relative to the size of the head, the dunnart has a large area 17 (V1) that occupies approximately one-third of the entire neocortex.

The striped possum is an arboreal marsupial that dwells in the tropical rain forests of Northern Queensland (Australia). This animal has an elongated fourth digit that is used to extract insects from holes in the tree bark, and specialized flaps that cover the lateral and frontal base of the incisors (J. E. Nelson, personal observation 1997). The cortex of the striped possum differs dramatically from that of the dunnart in that substantially less of the total amount of the neocortex is devoted to area 17, and the somatosensory cortex has expanded representations of the fourth digit, gums and tongue (Huffman et al 1999).

The most notable difference in the organization of marsupial neocortex compared to monotreme neocortex is the arrangement of cortical fields with respect to each other (Fig. 2). For instance, the primary visual area (Rosa et al 1999) is at the caudal pole of the cortex, and auditory cortex resides lateral to somatosensory cortex (Gates & Aitkin 1982, Aitkin et al 1986). In marsupials that have an expanded neocortex relative to their body size (e.g. the striped possum), more sensory fields have been observed, and more cortex is interposed between areas of the retained constellation (S1, A, V1, S2/PV, R, M).

Eutherian mammals

Although there is a dramatic variation in the size of eutherian mammal brains compared to metatherian (marsupials) and prototherian (monotremes) brains (Fig. 1), a number of components of organization are similar, despite over 150 million years of independent evolution. For instance, all eutherian mammals examined have S1, V1 and A1, as well as S2/PV, R and M. The general rostrocaudal organization of cortical fields is the same, with V1 residing most caudally, A1 located lateral and rostral to V1 and S1 located rostral to both (Fig. 2). The internal topographical organization of the retained cortical fields is also similar to that described in monotremes and marsupials. Like non-eutherian mammals, cortical fields in eutherians differ dramatically in the magnification of different portions of their sensory epithelium. For instance, most primates have a large magnification of the fovea in V1; this is particularly dramatic in a number of Old World monkeys. In monkeys that have a high degree of manual dexterity and tactile discriminatory abilities, such as those with glabrous hands with opposable thumbs, the representation of the glabrous hand assumes a large portion of all anterior parietal fields, including S1 (Nelson et al 1980, Kaas 1983). A similar cortical magnification of the glabrous hand representation in S1 has independently evolved in racoons (Welker & Seidenstein 1959, see Johnson 1990 for review). In racoons, there is a remarkable similarity in the structure of the hand compared to that of primates. The racoon's hand, unlike cats, is used extensively in fine tactile discriminations necessary for prey capture. Finally, like the mammals described previously, the amount of cortex devoted to a particular sensory system can vary dramatically, depending on the peripheral morphology of the animal, especially the size and density of receptor epithelium and the degree to which any specialized structure is used.

Two of the largest differences in neocortex of eutherian mammals compared to other mammals is an increased neocortical size in several lineages, and an increased cortical field number. For instance, a number of primates and cetaceans have a large cortical sheet that can be subdivided into a number of different cortical fields. In macaque monkeys, the number of visual fields alone has been estimated to be over 30 (Kaas 1997, Rosa 1997). The mechanisms by which cortical fields are added is not understood, and theories of cortical field addition are incomplete (e.g. Krubitzer 1995). However, it is clear that new cortical fields are not added hierarchically; rather, new fields are interspersed between the constellation of cortical fields that are present in all mammals, and therefore presumed to be evolutionarily older and retained from the common ancestor. Thus, with the addition of a new field to the retained plan, new interactions between afferent populations are established, and old connectional patterns while retained, are likely to have changed their relationships.

For instance, the ancestral mammal was likely to have possessed a network that consisted of connections between the lateral geniculate nucleus and V1, and between area V1 and extrastriate cortex immediately adjacent to it (V2). This basic network has been identified in a variety of species. Thus, marsupials and Old World monkeys share this component of visual cortical organization. However, new cortical fields have been added in primates, such as MT (the middle temporal visual area) and DL (the dorsolateral visual area). These new fields are interconnected with V1 as well as V2, and other extrastriate cortical fields and associated thalamic nuclei. Although the V1 to V2 connection has been maintained, the addition of new fields (and connections) to this old network is likely to have changed the existing network. While cortical fields and their connections may be homologous, it is unlikely that they are strictly analogous.

Similarities and differences

A survey of a wide variety of mammals indicates that there are common patterns of organization across species, which include a constellation of rostrocaudally and mediolaterally distributed cortical fields such as V1, S1, A1, R, M, PV/S2. While these fields appear to undergo large shifts in geographic location, and the amount of cortex allotted to any particular sensory system can vary dramatically, the general rostrocaudal organization is maintained, and thalamocortical and corticocortical connections share common patterns. There are several types of modifications that have been made to brains in evolution. Many of these changes take the same form, despite the fact that they have often evolved independently. Some of the modifications include: a change in size of the cortical sheet; a change in amount of cortex allotted to different sensory systems (Fig. 4); a change in the internal organization of a cortical field; shifts in cortical field location relative to other cortical fields; changes in connections of cortical fields; and module formation.

The evolution of cortical field development

These observations in extant mammals beg the question: are theories of cortical field specification within the life of the developing individual consistent with changes that are occurring in cortical fields in different lineages over time? Indeed, because a series of small changes to the developing nervous system are likely to account for much of the phenotypic variability in brain organization observed in extant mammals, theories of cortical development should accommodate the observations outlined throughout this chapter. Likewise, in order to understand the mechanisms of change, comparative neurobiologists must incorporate findings from developmental neurobiology in any theory of cortical field evolution. Thus, comparative studies combined with molecular techniques used to study the developing nervous system will allow us to answer several important questions regarding brain evolution. These include:

(1) How are rostrocaudal and mediolateral relationships between cortical fields maintained across species?
(2) What accounts for the large sensory domain shifts in cortical territory when cortical size is held constant?
(3) How does the cortical sheet increase in size and in the number of functional subdivisions?

The maintenance of thalamocortical afferent relationships with rostrocaudal coordinates of the cortex in all species examined suggests that there is likely to be some early specification in the developing cerebral cortex that helps to align incoming thalamocortical afferents (Fig. 5). A number of studies indicate that intrinsic patterning mechanisms within the telencephalon regulate regionalization of major subdivisions of the brain, independent of afferent input (e.g. Cohen-Tannoudji et al 1994, Levitt et al 1997, see Rubenstein et al 1999, Rubenstein 2000, this volume for review). For the neocortex in particular, one proposal is that the differential expression of regulatory genes such as *Emx2* and *Pax6* sets up a general rostrocaudal molecular gradient which controls the ordered growth of thalamic afferents to appropriate cortical locations (Gulisano et al 1996,

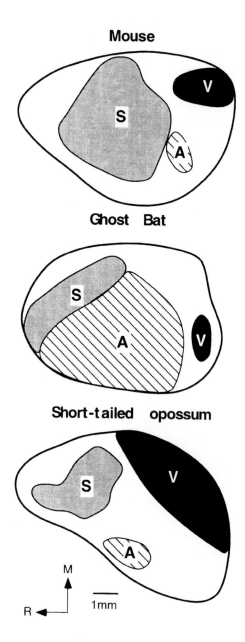

see Chenn et al 1997 for review). Such a mechanism could account for the stability of organization of homologous cortical fields observed in extant adult mammals. If such gradients exist to maintain relative thalamocortical relationships, how might they change in different lineages to induce differential occupation of sensory representations on the cortical sheet?

While it is possible that there may be some shift in gradients, particularly with changes in cortical sheet size (Fig. 5C, D), it is likely that other factors contribute to changes in the allotment of cortical territory in different species which have a similar sized neocortex, but have striking differences in peripheral morphology (e.g. mouse, ghost bat, short-tailed opossum; Fig. 4). The large shifts in afferent distribution observed in different lineages (Figs 4 and 6), in relation to changes in receptor density, distribution and type, suggest that major sensory domains are set up by the thalamus (O'Leary 1989, Schlaggar & O'Leary 1991), which reflects differences in the periphery (Kaas 2000, this volume). If cortical size is held constant, a genetically mediated change in the peripheral morphology can have resounding consequences for the organization of the entire neuroaxis (see Kaas 2000, this volume). Thus, while thalamocortical topographical relationships may be specified early via differential gene expression, precise visuotopic, somatotopic and cochleotopic distributions are likely to be driven by changes in the periphery, which in turn affects the organization of afferent target structures such as the brainstem and thalamus (Figs 5 and 6).

The final question of how the neocortex increases in size and number of cortical fields is not well understood. There are several recent suggestions about how the cortical sheet increases in size. One possibility is that decreased apoptosis (Kuida et al 1998, Rakic 2000, this volume) leads to an increase in the size of the cortical sheet. A second possibility is that an increase in the time over which cells proliferate in the ventricular zone can increase the size of the resulting cortical sheet exponentially (Rakic 1995, Kornack & Rakic 1998). It is more difficult to determine the possible mechanisms that contribute to

---

FIG. 4.   The organization of major sensory domains in distantly related species with a similar size neocortex. These species differ in their peripheral morphology and specialized behaviours associated with their expanded sensory systems. For instance, in the mouse, a large portion of the neocortex is devoted to processing somatic inputs, particularly from the vibrissae. The amount of cortex occupied by auditory and visual inputs is much smaller. The ghost bat is an echolocating microchiropteran bat that has devoted a substantial portion of its neocortex to processing auditory inputs. The visual system in the short-tailed opossum is well developed and occupies a large extent of the neocortex. The most notable feature of these brains is that if cortical sheet size is held constant, the cortical territory occupied by major sensory domains can change dramatically, depending on the sensory (peripheral) specialization of the animal. This suggests that specification of major sensory domains occurs later in cortical development and is dependant on activity from the periphery (see Kaas 2000, this volume). A, auditory cortex; M, motor cortex; R, rostral field; S, somatosensory cortex; V, visual cortex.

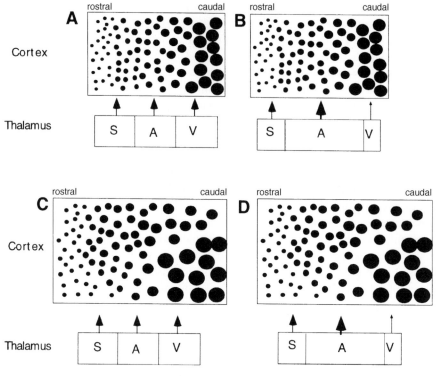

FIG. 5. Several possible ways in which sensory domain shifts might work in conjunction with molecular gradients to assign static thalamic afferent relationships in different mammals. (A) A high to low molecular gradient (e.g. *Emx2*) might work to align visual, auditory and somatosensory portions of the thalamus in an appropriate rostrocaudal fashion on the cortical sheet. Changes in major sensory representation may occur in the absence of changes to the molecular gradients that define rostrocaudal thalamocortical afferentation (B). Rather, a change in the size of the thalamic nuclei associated with a particular sensory system can result in a large change in the amount of cortex allotted to a particular sensory system (see Fig. 4). Thalamic changes are related to changes in the sensory epithelium. In a number of species, the size of the neocortex has increased. This may result in a small re-alignment of molecular gradients (C) in the absence of any dramatic changes to the thalamus. Finally, changes in both the cortical surface area and thalamic nuclei may account for differences in cortical allotment of different sensory systems (D), but this need not happen simultaneously in a particular lineage. It can be staggered in time. A, auditory inputs; S, somatosensory inputs; V, visual inputs. Large filled circles represent high concentrations and progressively smaller circles represent smaller concentrations of some molecule (taken in part from O'Leary et al 1999).

cortical field addition in different lineages. One possibility is that changes in activity patterns of incoming thalamic afferents generate new cortical fields (Krubitzer et al 1993, Krubitzer 1995). This might be accomplished simply by the addition of new cells to the developing thalamus (Fig. 6) which in turn

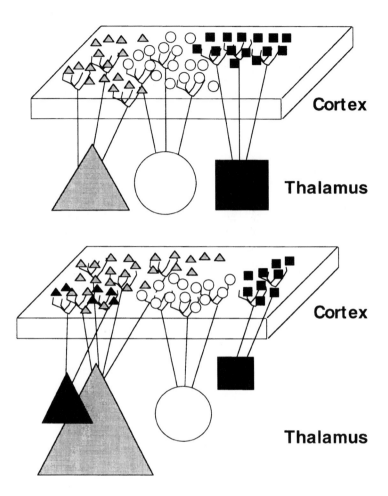

FIG. 6.   A representation of shifting afferents on the cortical sheet in different lineages in evolution. The hypothesized ancestral state is depicted at the top. In this figure, thalamic inputs from different nuclei (large square, circle and triangle) are distributed in some fashion on the cortical sheet. These afferent distributions have a rostrocaudal relationship to each other. In different lineages, the afferent distribution of inputs from particular thalamic nuclei is clearly altered (bottom) so that the relationships between afferents can change. While the rostrocaudal relationships appear to be largely maintained, there is a clear change in the amount of cortex particular afferents capture, and the relative position of afferents with respect to each other. In some species, new thalamic nuclei evolve, or new receptors evolve and generate modular organization within existing thalamic nuclei (small, dark triangle). The inputs from new receptor types (e.g. electrosensory receptors) are interspersed between existing inputs on the cortical sheet and form modules within a field. Modular organization of the neocortex is a common feature found across sensory systems and across mammals (see Krubitzer 1995).

projects to the cortex. This might also be accomplished by discorrelating old inputs from retained thalamic nuclei and combining them in novel ways to generate new cortical fields (Krubitzer 1998).

In conclusion, it has been proposed at this symposium that a few changes are likely to account for much of the phenotypic variability observed in the neocortex of extant mammals. I have tried to outline here some of the possible changes that might be occurring. These include genetically mediated changes in the size of the cortical sheet, changes in peripheral morphology, changes in receptor density, distribution and type, and changes in the thalamus that lead to discorrelations in activity of incoming afferents and/or addition of cells. It is improbable that all of these genetic changes are occurring simultaneously in any given lineage. Rather they are likely to be staggered across time in different lineages to ultimately produce the types of organizations observed in extant species.

Acknowledgements

I wish to thank Kelly Huffman, Jon Kaas, Dianna Kahn, Dennis O'Leary and Daniel Slutsky for helpful comments on this manuscript. This work was supported in part by the National Institutes of Health, the Whitehall Foundation and the MacArthur Foundation.

References

Aitkin LM, Irvine DRF, Nelson JE, Merzenich MM, Clarey JC 1986 Frequency representation in the auditory midbrain and forebrain of a marsupial, the northern native cat (*Dasyurus hallucatus*). Brain Behav Evol 29:17–28

Beck PD, Pospichal MW, Kaas JH 1996 Topography, architecture, and connections of somatosensory cortex in opossums: evidence for five somatosensory areas. J Comp Neurol 366:109–133

Chenn A, Braisted JE, McConnell SK, O'Leary DDM 1997 Development of the cerebral cortex: mechanisms controlling cell fate, laminar and areal patterning, and axonal connectivity. In: Cowan WM, Jessell TM, Zipursky SL (eds) Molecular and cellular approaches to neural development. Oxford University Press, New York, p 440–473

Cohen-Tannoudji M, Babinet C, Wassef M 1994 Early determination of a mouse somatosensory cortex marker. Nature 31:460–463

Gates GR, Aitkin LM 1982 Auditory cortex in the marsupial possum *Trichosurus vulpecula*. Hear Res 7:1–11

Gulisano M, Broccoli V, Pardini C, Boncinelli E 1996 *Emx1* and *Emx2* show different patterns of expression during proliferation and differentiation of the developing cerebral cortex in the mouse. Eur J Neurosci 8:1037–1050

Huffman KJ, Nelson J, Clarey J, Krubitzer L 1999 The organization of somatosensory cortex in three species of marsupials, *Dasyurus hallucatus*, *Dactylopsila trivirgata*, and *Monodelphis domestica*: neural correlates of morphological specializations. J Comp Neurol 403:5–32

Johnson JI 1990 Comparative development of somatic sensory cortex. In: Jones EG, Peters A (eds) Cerebral cortex, vol 8B: Comparative structure and evolution of cerebral cortex. Plenum Press, New York, p 335–449

Kaas JH 1983 What if anything is S1? Organization of the first somatosensory area of cortex. Physiol Rev 63:206–230

Kaas JH 1997 Theories of cortex organization in primates. In: Rockland KS, Kaas JH, Peters A (eds) Cerebral cortex, vol 12: Extrastriate cortex in primates. Plenum Press, New York, p 91–125

Kaas JH 2000 Organizing principles of sensory representations. In: Evolutionary developmental biology of the cerebral cortex. Wiley, Chichester (Novartis Found Symp 228) p 188–205

Killackey HP 1990 Neocortical expansion: an attempt toward relating phylogeny and ontogeny. J Cog Neurosci 2:1–17

Kirsch JAW, Mayer GC 1998 The platypus is not a rodent: DNA hybridization, amniote phylogeny and the palimpsest theory. Philos Trans R Soc Lond B Biol Sci 353:1221–1237

Kornack DR, Rakic P 1998 Changes in cell-cycle kinetics during the development and evolution of primate neocortex. Proc Natl Acad Sci USA 95:1242–1246

Krubitzer L 1995 The organization of neocortex in mammals: are species differences really so different? Trends Neurosci 18:408–417

Krubitzer L 1998 What can monotremes tell us about brain evolution? Philos Trans R Soc Lond B Biol Sci 353:1127–1146

Krubitzer L, Calford M, Schmidt L 1993 Connections of somatosensory cortex in megachiropeteran bats: the evolution of cortical fields in mammals. J Comp Neurol 327:473–506

Krubitzer L, Manger P, Pettigrew J, Calford M 1995 The organization of somatosensory cortex in monotremes: in search of the prototypical plan. J Comp Neurol 351:261–306

Kuida K, Haydar TF, Kuan C-Y et al 1998 Reduced apoptosis and cytochrome c-mediated activation in mice lacking caspase 9. Cell 94:325–337

Levitt P, Barbe MF, Eagleson KL 1997 Patterning and specification of the cerebral cortex. Annu Rev Neurosci 20:1–24

Manger P, Sum M, Szymanski M, Ridgway S, Krubitzer L 1998 Modular subdivisions of dolphin insular cortex: does evolutionary history repeat itself? J Cog Neurosci 10:153–156

Nelson RJ, Sur M, Felleman DJ, Kaas JH 1980 Representations of the body surface in postcentral parietal cortex of Macaca fascicularis. J Comp Neurol 192:611–643

O'Leary DDM 1989 Do cortical areas emerge from a protocortex? Trends Neurosci 12:401–406

O'Leary DDM, Yates PA, McLaughlin T 1999 Molecular development of sensory maps: representing sights and smells in the brain. Cell 96:255–269

Rakic P 1995 A small step for the cell, a giant leap for mankind: a hypothesis of neocortical expansion during evolution. Trends Neurosci 18:383–388

Rakic P 2000 Radial unit hypothesis of neocortical expansion. In: Evolutionary developmental biology of the cerebral cortex. Wiley, Chichester (Novartis Found Symp 228) p 30–45

Rosa MGP 1997 Visuotopic organization of primate extrastriate cortex. In: Rockland KS, Kaas JH, Peters A (eds) Cerebral cortex, vol 12: Extrastriate cortex in primates. Plenum Press, New York, p 127–203

Rosa MGP, Krubitzer L, Molnár Z, Nelson JE 1999 Organization of visual cortex in the northern quoll, Dasyurus hallucatus: evidence for a homologue of the second visual area in marsupials. Eur J Neurosci 11:907–915

Rowe M 1990 Organization of the cerebral cortex in monotremes and marsupials. In: Jones EG, Peters A (eds) Cerebral cortex, vol 8b: Comparative structure and evolution of the cerebral cortex. Plenum Press, New York, p 263–334

Rubenstein JLR 2000 Intrinsic and extrinsic control of cortical development. In: Evolutionary developmental biology of the cerebral cortex. Wiley, Chichester (Novartis Found Symp 228) p 67–82

Rubenstein JLR, Anderson S, Shi L, Miyashita-Lin E, Bulfone A, Hevner R 1999 Genetic control of cortical regionalization and connectivity. Cereb Cortex 9:524–532

Schlagger BL, O'Leary DDM 1991 Potential of visual cortex to develop an array of functional units unique to somatosensory cortex. Science 252:1556–1560

Sousa APB, Gattass R, Oswaldo-Cruz E 1978 The projection of the opossum's visual field on the cerebral cortex. J Comp Neurol 177:569–587

Stephan H, Baron G, Frahm HD 1988 Comparative size of brains and brain components. Comp Primate Biol 4:1–38

Welker WI, Seidenstein S 1959 Somatic sensory representation in the cerebral cortex of the racoon (*Procyon lotor*). J Comp Neurol 111:469–501

Westerman M, Edwards D 1992 DNA hybridization and the phylogeny of monotremes. In: Augee ML (ed) Platypus and echidna. The Royal Zoological Society of New South Wales, Mosman, p 28–34

DISCUSSION

Karten: What is the most important take-home message about your work?

Krubitzer: We need to re-evaluate what a cortical field is, and not just think of it as a homogenous structure. It is not as straightforward as this. A new cortical field is just a new pattern of connection interactions.

Purves: You have demonstrated that cortical fields often comprise iterated units of some sort, but is it fair to conclude that all of these so-called modules represent the same kind of entity? I'm thinking in particular about barrels and ocular dominance columns. Can barrels, which represent particular peripheral receptors and that arise in the cortex more or less independently of the function of the system, be compared with ocular dominance columns, which have no peripheral correlate and seem to be much more dependent on activity?

Krubitzer: The similarity between the two is that cortex is segregating inputs from the two eyes and from the separate barrels. In addition, if you look at the sizes of modules across animals with different sized brains, you find that even if the size of the cortex varies by a factor of a few thousand, the module size only varies by a factor of two. Therefore, the cortex is not only segregating inputs, but it is maintaining a constant module size. There must be a number of competing factors that account for this restricted size, and people have argued that there's a selection for minimal connection length.

Pettigrew: I would like to add that we have been working on the platypus, and it has stripes in its primary somatosensory cortex that are like the ocular dominance columns in monkey striate cortex. The barrel fields segregate inputs from whiskers, the ocular dominance columns segregate input from the two eyes, and in the case of the platypus there are mechanoreceptors that detect mechanical vibrations in the water and electroreceptors that detect electrical information in the water. It turns out that the two sets of stripes in the platypus do what I think (we can't prove it yet), they will carry out a sophisticated analysis comparing two different sensory arrays for small differences that will give distance, just as the ocular dominance columns do in monkeys. The ocular dominance columns enable the monkey to interdigitate the two representations of the two eyes, so it can have binocular vision. However, there is a paradox because, although there is segregation of inputs into barrels, in the case of the barrels you don't want information from one whisker, you want all the information from all the whiskers, in order to obtain information on the shape of a bundle of food, for example. Therefore, at some level all this information has to be put together—a tension between segregation of inputs and fusion. Another example of this occurs in the auditory system of the owl, where the same afferent is split into two pathways, i.e. for time analysis and for intensity analysis. These give rise to two representations that deal with time and intensity information. The two, when finally put back together, give rise to a map of auditory space that could not have been generated by mixing the inputs early on. In the case of the platypus, the two sets of stripes analyse prey, e.g. if a shrimp flicks its tail, the electrical stimulus arrives instantaneously, whereas the mechanical stimulus takes much longer to travel through the water (Pettigrew et al 1998). Therefore, the neurons above layer IV, where these two stripes integrate, give rise to information about how far away the stimulus is. In the visual system, the advantage of mixing information from the two columns is that it gives rise to stereopsis. Therefore, we have to realize that there is segregation, but that it is also integrated in the layers above layer IV to give rise to a novel function.

Rubenstein: Is the segregation of the mechanoreceptors and electroreceptors stripes present at birth?

Pettigrew: The trouble with studying the platypus is that it is 60% politics and 40% science, so the number of animals we can work on is small and our developmental data are limited. We do know that electroreceptors develop first. It's possible that the young are using these to find the mother's milk.

Rakic: The fundamental difference between ocular dominance columns and both mechanoreceptors and electroreceptors is in the ocular dominance columns there is segregation of inputs from two sides subserving the same function, whereas for mechanoreceptors and electroreceptors there is segregation for two different functions. A more comparable example would be the development of the M and P subsystems in the primate visual system. These two subsystems, which subserve different functions, develop early and independently by the molecular cues (Meissirel et al 1997). In contrast, ocular dominance columns are determined by competition between functionally identical inputs from the two sides (Rakic 1981).

Pettigrew: But in barrels we are talking about segregation or fusion of inputs from the same whisker, the same modality. There are two images of the outside world and the animal has to analyse those, as well as it can, before it mixes them.

If they are mixed too early, the fine information needed to make this integration possible is lost.

Hunt: You implied that it is necessary to keep representations separate to get the maximum amount of information. If you didn't do this would you get any information at all?

Pettigrew: Yes. There are some marsupials that bring the eyes together in the thalamus. They don't have stereopsis, but it does mean that they can increase the signal to noise ratio. But the point is that there are definite advantages of mixing the two representations.

O'Leary: Leah Krubitzer showed some beautiful examples where the entire cortex was the same size, but the size, and sometimes absolute location within the cortex, of relative primary sensory areas differed. To explain how area-specific thalamocortical targeting occurs during development to generate those types of patterns, I imagine there would need to be proportional changes in the diencephalon, perhaps in terms of the size of the various primary sensory nuclei as well as their relative locations. Have you looked at whether these correlations exist?

Krubitzer: Yes, In the ventroposterior nucleus of the platypus, trigeminal input dominates almost the entire thalamus.

O'Leary: What do you observe in the thalamus of the platypus where V1 is shifted more rostral and medial relative to S1, compared to a species such as a rat? Is there a correlation between the relationships of V1 and S1 and the relative positioning of the lateral geniculate nucleus and the ventroposterior nucleus?

Krubitzer: It is difficult to identify where the lateral geniculate nucleus is in the platypus. Regidor & Divac (1987) have placed it ventral to the ventroposterior nucleus, whereas Campbell & Hayhow (1971) and Ulinski (1984) have suggested that it is a small wedge at the lateral portion of the thalamus, However, it hasn't shifted dramatically. Even though there is a shift in these cortical fields, the visual cortex is not in front of somatosensory cortex, and in our developmental studies, there is not a complete reversal of the positions of visual and somatosensory cortex.

Levitt: Wouldn't this suggest that some sort of organizing centre is present?

Krubitzer: Yes. After listening to some of the presentations at this symposium, and seeing some of the borders, I have come to the conclusion that there must be.

Herrup: It would be worth tracing the tangentially migrating cells from the ganglionic eminence to find out if they play a role in parcelling out these fields. The cells of neocortex are generated in a strict order with respect to location and phenotype, and their fate is governed in part by when they were born. If there were similar temporal changes in the quality of cells coming out of the ganglionic eminence, you might conclude that their fates were similarly linked to their birth date.

Levitt: But the evidence suggests that there is lineage inheritance, so it's the glutaminergic projection neurons that are specified. The γ-aminobutyric acid (GABA)ergic neurons may not need specific field information, perhaps.

Herrup: If you did birth date studies of the GABAergic cells, you might find that all the early-born cells are posterior. The prediction would be, if you engineered a smaller cortex by taking out a section from the anterior region, and that operation blocked the migratory path, you would generate a dysfunctional cortex. The reason it works is that you take out a posterior portion and the cells are all going that way.

Karten: Another point that stands out is that what is reflecting is the changing periphery more than anything else.

Herrup: But how could you also change the balance?

Karten: The question is, does coding occur at one level or at all levels, e.g. the size of the nose, the size of the eye, the numbers of ganglion cells and the nuclei. Events don't evolve singularly, what we see is an end picture, so if you only focus on the brain without keeping in mind that these are all derivative factors, you may run the risk of falsely isolating one phenomenon.

Reiner: I was interested in Leah Krubitzer's data on map relocation and cortical ablations. What are the implications of these results for the specificity of handshake hypothesis? Certain cortical afferents have to shake hands with certain thalamic afferents in order for them to end up in the correct place. In this case the wrong ones shake hands, but an appropriate field still forms in more or less the appropriate place relative to other fields. This seems to argue for some kind of ordered ingrowth notion rather than for a specificity of handshake notion. It also seems that Leah Krubitzer's results have implications for the notion that gradients of molecules specify where particular afferents will go. Her results suggest that it is not the absolute level of the molecule that is critical, but rather the relative levels.

Molnár: There is evidence for plasticity after the actual handshake, once the afferents go through the internal capsule (see discussion after Molnár 2000, this volume). The key to explaining this, and the large amount of relocation of thalamocortical projection, is that early activity must be involved.

Levitt: When you make those lesions, there is a retrograde effect, but you don't know when this occurs. It must be impacting on the thalamus, and if you remove cortex, you also remove thalamus. During development, those neurons die quickly, and unless you follow this closely it's difficult to know what the initial relationship was between cortex and thalamus, and how soon the thalamus adapts to the lesion.

Molnár: Pat, you are referring to early experiments in newborn rats in which lesions are made in the visual cortex (Cunningham et al 1987). In these rats, the lateral geniculate nucleus degenerates after the cortical lesion. The stage we used for our experiments in *Monodelphis*, however, corresponds to embryonic day (E) 11 in the rat (Molnár et al 1998), which may explain the difference. Perhaps at this early stage thalamus does not depend on cortical factors for its survival and this would provide enough time for the thalamocortical projections to distribute over the smaller cortex. When the dependency of the thalamic neurons sets in, the matching between thalamus and reduced cortex occurs on a general scale in *Monodelphis* rather than only in the nuclei originally destined to the lesioned cortical site. In our *Monodelphis* experiments, the thalamus reduces in size in proportion to the size of the cortical lesion, and the lateral geniculate nucleus remains present, although smaller. The comparison of the study of Cunningham et al (1987) and ours suggests that if lesions are made later, the projections are specified and they have no means of re-specifying themselves, but if the lesion is made at an age equivalent to E11, then they can still substantially rearrange and it's a completely different type of experiment paradigm.

Puelles: Another way to see constraint in the way the different fields have evolved, is that not only is the overall topology of different sensory, motor, limbic or associative areas maintained, but also the new areas for each modality are connected to the others, i.e. functionally related areas all remain close neighbours. This speaks of an invisible boundary isolating the local common modality from other modalities during development. Therefore, new visual fields may appear because the individual pre-existent fields allow new positions within the common boundary to be occupied by additional maps, rather than extending themselves (this assumes that differences in the number of cortical maps really exist and there is no problem with map detection procedures). This means to me that the new map-generating mechanism has a morphogenetic basis, i.e. it is only possible to have new visual field representations in the visual portion of the overall topology, but the total number of maps can increase possibly due either to overall increase in size of the cortex, or to miniaturization of the essential modules without accompanying size increase. All this implies parameter changes within the all-enclosing field and a resulting novel equilibrium state of the system. Apart from genetic constraint due to early specification of precursor populations in the diverse cell sources, there may also be some additional constraint imposed by the way the fibres come from the thalamus and fasciculate/defasciculate in the subcortical white matter.

Krubitzer: So you don't evolve a new visual field within the somatosensory cortex. New visual areas always appear adjacent to or intermingled with old visual areas. The way we tend to think about processing implies that cortical fields are added hierarchically, but this is not the case. They are inserted into existing networks, and may often evolve from primary fields, which makes primary fields difficult to identify as homologous. The geometric relationship of thalamocortical afferents to each other is important, and if a new type of visual input evolves, it will stick close to things that it's correlated with in terms of activity patterns.

Goffinet: Several years ago Divac & Oberg (1990) reported that the echidna had a large thalamic MD and a large prefrontal cortex. Do you agree with this?

Krubitzer: The notion that echidnas have a large prefrontal cortex is incorrect, because when you make inferences about evolution you have to think of the most parsimonious explanation that accounts for a given structure. If this notion were correct, you would have to argue that there were many transformations in between monotremes and humans, and this is unlikely. What has probably occurred is that there has been an independent expansion of orbital frontal cortex, and the large frontal cortex in echidnas is probably related to the olfactory system. In terms of their MD size, it really isn't that large.

Goffinet: So, this frontal lobe would be olfactory?

Krubitzer: It probably is olfactory. If you visually inspect the brain of the animal you can understand a lot about it. This echidna's brain has a large olfactory bulb that has many fissures. When you flatten the cortex, you see that a large portion of the cortex is occupied by piriform cortex, compared to sensory cortex.

Puelles: What is the size of the claustrum in echidnas?

Krubitzer: We haven't looked at that.

Butler: Divac et al (1987) have reported that echidnas do not have a claustrum, so this structure may be unique to marsupial and placental mammals.

Karten: Does anyone know what the claustrum is doing physiologically?

Pettigrew: It may represent a link between behaviour and the re-jigging of visual cortical circuitry.

Karten: How different is the cytoarchitecture in the claustrum compared to the isocortex?

Puelles: It is rather homogeneous at any transverse section plane, although there are distinct areas and subnuclei along its considerable rostrocaudal extent (from olfactory nucleus and orbital cortex, anteriorly, to amygdala, caudally); these express different immunocytochemical markers and show different distributions of transmitter receptors or neuropeptide receptors. A large part of the claustrum is connected reciprocally point to point with the whole isocortex, which creates claustral areas with presumed predominant function related to vision, somesthesia, hearing, olfaction, etc. Thalamic afferents have been described from the posterior and intralaminar nuclear complexes in the dorsal thalamus.

Karten: Have any behavioural deficits been associated with lesions in the claustrum?

Puelles: The problem is that experiments placing tracer injections and making lesions in the claustrum are difficult to analyse, because you often cannot be sure whether the connections you see belong to the claustrum itself, as compared to the closely neighbouring insular or piriform cortexes, or the striatum, due to the problem of fibres of passage. There are also few scientists who work specifically on the claustrum. However, mammals with scarce and relatively primitive isocortex, such as the echidna and platypus, are possibly of choice for investigating this daunting component of the pallium. I would be surprised if it is not present and rather massive in these animals. After all, it must have evolved from some earlier primordium in stem amniotes. Divergent evolution of the mammalian cortex from the reptilian cortex probably cannot be understood completely without explaining the joint origin of the cortex, the claustrum and the pallial parts of the amygdala (Striedter 1997, 1998, Puelles et al 2000). Our relative inconclusiveness (or lack of interest) about relevant field homologies (see discussion elsewhere in this book on the sauropsidian dorsal ventricular ridge) handicaps drawing uncontested conclusions from studies restricted to comparing isolated cell populations.

References

Campbell CB, Hayhow WR 1971 Primary optic pathways in the echidna, *Tachyglossus aculeatus*: an experimental degeneration study. J Comp Neurol 143:119–136

Cunningham TJ, Haum F, Chantler PD 1987 Diffusable proteins prolong survival of dorsal lateral geniculate neurons following occipital cortex lesions in newborn rats. Dev Brain Res 37:133–141

Divac I, Oberg RGE 1990 Prefrontal cortex: the name and the thing. In: Scherdtfeger WK, Germroth P (eds) The forebrain in nonmammals. Springer-Verlag, Berlin, p 213–220

Divac I, Holst M-C, Nelson J, McKenzie JS 1987 Afferents of the frontal cortex in the echidna (*Tachyglossus aculeatus*). Indication of an outstandingly large prefrontal area. Brain Behav Evol 30:303–320

Meissirel C, Wikler KC, Chalupa LM, Rakic P 1997 Early divergence of magnocellular and parvocellular functional subsystems in the embryonic primate visual system. Proc Natl Acad Sci USA 94:5900–5905

Molnár Z 2000 Conserved developmental algorithms during thalamocortical circuit formation in mammals and reptiles. In: Evolutionary developmental biology of the cerebral cortex. Wiley, Chichester (Novartis Found Symp 228) p 148–172

Molnár Z, Knott GW, Blakemore C, Saunders NR 1998 Development of thalamocortical projections in the South American grey short-tailed opossum (*Monodelphis domestica*). J Comp Neurol 398:491–514

Pettigrew JD, Manger P, Fine SLB 1998 The sensory world of the platypus. Philos Trans R Soc Lond B Biol Sci 353:1199–1210

Puelles L, Kuwana E, Puelles E et al 2000 Pallial and subpallial derivatives in the embryonic chick and mouse telencephalon. submitted

Rakic P 1981 Development of visual centers in the primate brain depends on binocular competition before birth. Science 214:928–931

Regidor J, Divac I 1987 Architectonics of the thalamus in the echidna (*Tachyglossus aculeatus*): search for the mediodorsal nucleus. Brain Behav Evol 30:328–341

Striedter GF 1997 The telencephalon of tetrapods in evolution. Brain Behav Evol 49:179–213

Striedter GF 1998 Progress in the study of brain evolution: from speculative theories to testable hypotheses. Anat Rec 253:105–112

Ulinski PS 1984 Thalamic projections to the somatosensory cortex of the echidna, *Tachyglossus aculeatus*. J Comp Neurol 229:153–170

Developmental plasticity: to preserve the individual or to create a new species?

Egbert Welker

Institut de Biologie Cellulaire et de Morphologie, Université de Lausanne, Rue du Bugnon 9, 1005 Lausanne, Switzerland

Abstract. The cerebral cortex has an amazing capacity to adjust its organization in response to perturbations of its normal development. This developmental plasticity can be considered to have, as its ultimate goal, the preservation of an 'intact' individual, capable of integrating sensory information to generate an adequate behavioural response. The mechanisms underlying developmental plasticity, however, can also be considered of importance to generate variability among individuals of the same species and, as such, create the platform for evolution to occur. Here I describe three experiments that alter the configuration of the somatosensory cortex of the mouse. The first is based on the removal of whisker follicles neonatally and demonstrates that the formation of barrels is dependent of the presence of follicles. The second is based on results of selective inbreeding for the number of sensory organs (whisker follicles) and illustrates the strong tendency during the period of developmental plasticity to preserve the internal organization of the cerebral cortex. The third experiment is based on a mutation that affects the formation of barrels and, as a consequence, alters cortical processing of sensory information. This mutation can be considered to have resulted in an evolutionary deviation.

2000 Evolutionary developmental biology of the cerebral cortex. Wiley, Chichester (Novartis Foundation Symposium 228) p 227–239

Genetic variation within a population is a requirement for evolution to occur. It forms the basis for phenotypic variations that, in the situation of a modified selective pressure, could result in divergence within the population, eventually resulting in the formation of a new species. This theoretical framework pertains to phenotypic variations of the entire body, that, from a 'neuronal' point of view, is composed of two compartments: the central nervous system (CNS) and the periphery. This simplified version of the body plan raises the question of how genetic variation in one compartment gives rise to structural and functional modifications in the other compartment. Or, put differently, could phenotypic variations within one compartment be the consequence of genetic modifications that took place in the other compartment?

Neurobiologists have studied the factors involved in the establishment of a correspondence between the periphery and the CNS in a variety of systems throughout the animal kingdom. In this context, sensory systems have the simplest relationship. They carry information from the periphery towards, and subsequentially, within the CNS. In these systems, the periphery is the site where the sensory stimulus activates peripheral neurons that, in the case of mammalian species, terminate with their other extremity inside the CNS. The image that the animal creates depends on the signal the CNS receives from the peripheral sensory sheet and its subsequent processing within the CNS. It forms the basis for the animal's perception of the outside world. The establishment of the connectivity between the periphery and the CNS during development is to a great extent determining what the animal may experience later in its life.

During development, the nervous system has a degree of plasticity, i.e. it has the capacity to adapt its organization as a response to a perturbation of its normal development. This aspect of the nervous systems is widely studied in the context of the establishment of a correspondence between the sensory periphery and its representation within the CNS. Plasticity has been demonstrated to occur in sensory systems of all species studied and should be considered to be a highly preserved trait in the development of the CNS. The importance is obvious, e.g. in the case of an extra finger, the spatiotemporal integration of the input from the modified hand (sensory periphery) can only be established if the extra element has a representation in a topological 'correct' site within the central representation of the hand. The central representation of the extra finger should be in an appropriate spatial relationship with those of the normal fingers. In this example, neuronal plasticity allows the central representation to modify its organization in such a manner that the individual, later in life, can make sense out of the information coming from their modified peripheral sensory sheet.

The cortical representation of the mystacial whisker follicles in the mouse

The whisker-to-barrel pathway of the mouse will be used to illustrate a few of the concepts introduced above. In this species, it forms the part of the somatosensory system that processes information from the whisker follicles. These sensory organs are embedded at various sites of the skin. One group of whisker follicles is present at the snout of the animal and are called the mystacial whiskers. Together with the intervening skin these whiskers form the whiskerpad. Mystacial whisker follicles are distributed in five horizontal rows (named A–E; Fig. 1A); four straddling follicles $(\alpha–\delta)$ are placed caudally of the five rows. The neurobiological interest in whisker follicles was raised by the observation that the representation of individual whisker follicles, at the level of the somatosensory cortex, could be visualized using ordinary histological methods (Woolsey & Van der Loos 1970). At the level of layer IV, the neuronal somata are distributed such that they form cell body-dense rings that surround areas where cell bodies are relatively scarce. The three-dimensional organization of these multineuronal units is the origin of their name, barrels. The distribution of barrels, as seen in sections made tangential to the pial surface of the somatosensory cortex, reflects the organization of the mystacial whisker follicles. There are five rows of barrels (A–E) with four straddling elements (α–δ). This part of the somatosensory cortex is called barrel cortex. Since the first description of barrels by Woolsey & Van der Loos (1970) the one-to-one correspondence between a whisker follicle and a cortical barrel was demonstrated in neurophysiological and deoxyglucose studies (Welker 1971, Simons 1978, Melzer et al 1985, Nussbaumer & Van der Loos 1985, Chmielowska et al 1986, Armstrong-James & Fox 1987).

In the following sections I will describe three sets of experiments that affect the development of barrels in different ways.

Neonatal whisker removal

In mice, barrels form at postnatal day (P) 4 (Rice & Van der Loos 1977). In the case of whisker follicle removal before this age, barrels do not form. This was first demonstrated by Van der Loos & Woolsey (1973) using an experimental paradigm which varied in the number of whisker follicles removed. Removal of a single row of follicles was subsequently used to show that lesions made at various postnatal ages produce different alterations in the cortex (Jeanmonod et al 1981). Figure 1B illustrates the consequence of a peripheral intervention carried out at the second postnatal day and is characterized by the enlargment of the representation of the neighbouring row of (untouched) whisker follicles. Interestingly, the innervation density of these spared follicles is *not* altered, as compared to intact animals (Welker & Van der Loos 1986a). In other words, the enlargement of the central representation of the neighbouring follicles is not a mere reflection of an increased innervation of the neighbouring follicles. The rules underlying the enlargement of the neighbouring row of barrels seems, therefore, to reside within the CNS. Schlaggar et al (1993) showed that postsynaptic activation via the NMDA receptor is involved in the modification of the cortical representation.

Other studies revealed that the cortical modifications are accompanied by similar rearrangements in the subcortical stations of the somatosensory pathway (Belford & Killackey 1980, Durham & Woolsey 1984). Lesions made after P4 do not modify the cytoarchitecture of the barrel cortex. This led to the definition of a critical period for the development of barrels (Jeanmonod et al 1981). Lesions made at later ages, however, do modify intracortical connectivity (McCasland et al 1992) and the functional organization of the whisker representation (e.g. Melzer &

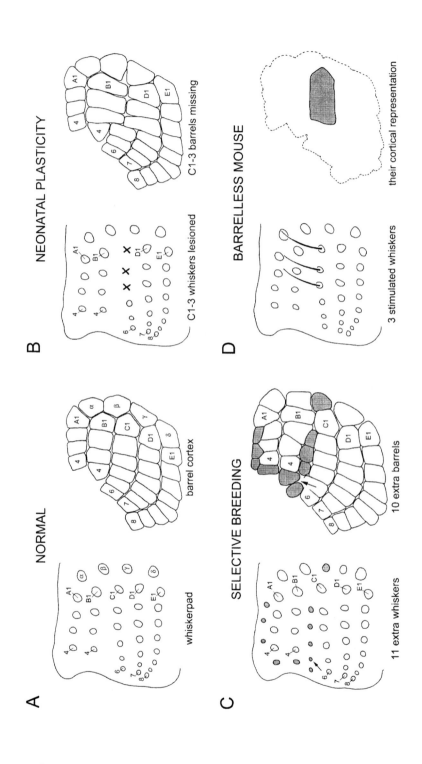

A NORMAL

whiskerpad · barrel cortex

B NEONATAL PLASTICITY

C1-3 whiskers lesioned · C1-3 barrels missing

C SELECTIVE BREEDING

11 extra whiskers · 10 extra barrels

D BARRELLESS MOUSE

3 stimulated whiskers · their cortical representation

Smith 1998, Bronchti et al 1999), but do not change the cytoarchitectural structure
of the barrel.

Selective breeding for variations in number of mysticial whisker follicles

The observation of a supernumerary whisker follicle in a mouse was the starting
point of a detailed screening of a population of 597 mice from an outbred stock. In
about 50% of these mice supernumerary (extra) whisker follicles were detected
(Van der Loos et al 1984). They were all confined to the part of the skin that
corresponds, during embryogenesis, to the zones of mergence between the three
embryonic facial folds that together form the upper lip of the mouse. A number of
these mice were used for a selective breeding program. The results demonstrated
that there is a genetic basis for the occurrence of supernumerary whiskers. The
breeding program established a number of different lines of mice (Van der Loos
et al 1986), each characterized by a defined pattern of whisker follicles: a line in
which the incidence of supernumerary whiskers is absent (normal mice); lines
that are characterized by a constant number of supernumerary whiskers
occurring at defined sites on the whiskerpad; and other lines, in which the mean
number of supernumerary whisker follicles per individual increased over the

---

FIG. 1.   Illustration of the pattern of mystacial whisker follicles and their representation in the
somatosensory cortex of normal mice (A) and in three experimental conditions: (B) after whisker
follicle lesion at the first postnatal day; (C) in mice selectively bred for a maximal number of
supernumerary whisker follicles; and (D) in the strain of *barrelless* mice. (A) The distribution of
the mystacial whisker follicles are displayed to the left where they are represented by open circles.
The first whisker follicle in each horizontal row is labelled by a letter (A–E); the straddling
follicles are labelled by Greek letters (α–δ). The introduction of these labels helps to highlight
the correspondence between the distribution of whisker follicles and their representation at the
level of the somatosensory cortex (barrel cortex), where the outline of barrels is drawn first. (B)
The effect of the lesion of the first three follicles of row C (crosses indicate the site of the
peripheral intervention) at the second postnatal day. Note that the three corresponding
barrels of row C are missing and that the neighbouring barrels of rows B and D are enlarged
and fill the 'vacant' position in the barrel cortex. (C) Eleven supernumerary follicles at the
whiskerpad of a mouse from the line bred for a maximal number of mystacial whisker follicles.
The supernumerary follicles are rendered grey in the display of the whiskerpad. Ten of them have
a corresponding barrel in the cortex, one, indicated by an arrow, is not represented by a barrel:
the arrow in the barrel cortex points to the site where it's barrel was expected to occur. (D)
Topological organization of the whisker representation in the somatosensory cortex of the
barrelless mouse as illustrated here by the outcome of a deoxyglucose study. During the
experiment the animal was exploring a stimulus-rich cage with only three whiskers in row
C—the others had been clipped just prior to the deoxyglucose injection. The grey area in the
cortex represents the site of the stimulus-dependent deoxyglucose uptake. The stippled line
indicates the outline of the whisker representation in these mice as observed in the Nissl-
stained sections, in which individual barrels could not be identified. Note that the
metabolic activity is confined to a 'normal' part of S1.

subsequent generations, resulting in the recent record of a mouse possessing up to 43 supernumerary whiskers.

Whisker follicles are sensory organs that develop at the level of the skin and receive their sensory innervation from neurons in the Gasserian ganglion. The question was formulated: how do the periphery and central nervous system respond to an increased number of whisker follicles? A quantitative analysis in several of the lines described above revealed that supernumarary whiskers are not only innervated, but it does not occur at the expense of the innervation of the 'standard' whisker follicles. On the contrary, in all lines with supernumerary whiskers studied, the innervation of the standard whisker follicles was found to be increased in comparison to the innervation of these follicles in normal mice (Welker & Van der Loos 1986b).

Turning to the CNS, we observed, in most cases, the supernumerary whisker follicles were represented by supernumerary barrels. These barrels were most often localized at the topologically corresponding site of the cortical face representation (Fig. 1C). In the mouse line bred for a maximal number of supernumerary whisker follicles we found that the presence of supernumerary barrels led to an increased area of cortical whisker representation. This increase was about 10% of the entire representation (Welker & Van der Loos 1986b).

Interestingly, in a few cases the supernumerary whisker follicle was *not* represented by a cortical barrel (see arrows in Fig. 1C). Comparing the innervation density of these follicles with those that were represented by a barrel, led to the notion that to induce the formation of a barrel the peripheral whisker follicle needs to be innervated by at least 30 myelinated nerve fibres (Welker & Van der Loos 1986b).

This set of observations on the results of a selective breeding programme can be placed in the context of evolution in the following manner:

- The cerebral cortex has the capacity to create space for the representation of 'new' sensory organs.
- The induction of cortical representation needs a peripheral signal above a certain threshold that is most likely related to a level of neuronal activity.
- Plasticity of the peripheral and central nervous system forms a neuronal basis for interindividual variability and, therefore, fulfills a requirement for evolution to occur.

barrelless mice — a spontaneously occurring mutant

Among the line of normal mice (see above) we accidently discovered animals that lacked barrels. Upon checking the breeding records we identified that these mice had a common ancestor. Subsequent screening led to the identification of mice that lack barrels on a genetic basis. Accordingly, this strain was named *barrelless*. Mice of this strain are homozygous for the *brl*-mutation that was mapped to chromosome 11 (Welker et al 1996). It was further shown that the mutation had disrupted the gene coding for the enzyme adenylyl cyclase 1 (Abdel-Majid et al 1998). Besides the lack of barrel formation, other aspects of the differentiation of the cerebral cortex look normal. Notably, the cortical layers are present and of equal size in normal and *barrelless* animals. In both strains of mice, the thalamocortical projection conveying whisker information to the somatosensory cortex terminates in layers IV and VI. The difference between the normal and the *barrelless* mouse was found to be in the tangential organization of the thalamocortical axons. In the normal mouse, these axons segregate in individual barrel domains, whereas in the *barrelless* animal such segregation is not present in the adult somatosensory cortex. Interestingly, in the subcortical station of the whisker-to-barrel pathway (brainstem and thalamus) the whisker-related patterns are present.

Using deoxyglucose autoradiography we demonstrated that the cortical whisker representation in the *barrelless* mouse is topologically organized in a similar manner to barrel-rich animals (illustrated in Fig. 1D). So the question arose: does the lack of barrels have a functional consequence on the processing of whisker information? The analysis of the response properties of single units in the urethane anaesthetized mouse provided an answer. In normal mice, cortical neurons respond to the stimulation of a principal whisker (i.e. the whisker that corresponds to the barrel in which the cell body is situated), as well as to the stimulation of surrounding whiskers. The difference in the cortical activation between these two classes of sensory input is that the response to principal whisker stimulation is of shorter latency than that of surround whiskers and also has a larger magnitude (Welker et al 1993). In *barrelless* mice, this difference is not maintained and the majority of cortical cells respond with equal latency and magnitude to the stimulation of two different whiskers (Welker et al 1996). This loss of spatiotemporal discrimination of the whisker input should have a consequence for the behavioural use of the mystacial vibrissae in the *barrelless* mouse. In an evolutionary perspective, the *barrelless* mutation has created a mouse with a sensory impairment that, in a natural setting, could have dramatic consequences on its survival.

In conclusion, these studies demonstrate three different sources of interindividual variability in the whisker-to-barrel pathway of the mouse. It demonstrates the striking tendency of the nervous system to adapt its organization to a modified pattern of peripheral, sensory organs. This adaptability can be interpreted as an expression of neuronal plasticity through which the individual preserves the behavioural use of a sensory modality. In case of a modified selective pressure, however, this interindividual variability may have evolutionary consequences.

Acknowledgements

This article is dedicated in memory of Hendrik Van der Loos, who introduced me in the barrel world and with whom I share its content. I like to thank Dr G. R. Bock and Ms K. Webb from the Novartis Foundation for their patience during the preparation of the manuscript; Dr G. W. Knott for critical reading of the manuscript; Mr E. Bernardi for the help with the illustrations. This work is supported by the Swiss National Science Foundation grant 3100-51036.97.

References

Abdel-Majid RM, Leong WL, Schalkwyk LC et al 1998 Loss of adenylyl cyclase I activity disrupts patterning of mouse somatosensory cortex. Nat Genet 19:289–291

Armstrong-James M, Fox K 1987 Spatiotemporal convergence and divergence in the rat S1 'barrel' cortex. J Comp Neurol 263:265–281

Belford GR, Killackey HP 1980 The sensitive period in the development of the trigeminal system of the neonatal rat. J Comp Neurol 193:335–350

Bronchti G, Corthésy ME, Welker E 1999 Partial denervation of the whiskerpad in adult mice: altered patterns of metabolic activity in barrel cortex. Eur J Neurosci 11:2847–2855

Chmielowska J, Kossut M, Chmielowski M 1986 Single vibrissal cortical column in the mouse labeled with 2-deoxyglucose. Exp Brain Res 63:607–619

Durham D, Woolsey TA 1984 Effects of neonatal whisker lesions on mouse central trigeminal pathways. J Comp Neurol 223:424–447

Jeanmonod D, Rice FL, Van der Loos H 1981 Mouse somatosensory cortex: alterations in the barrelfield following receptor injury at different early postnatal ages. Neuroscience 6:1503–1535

McCasland JS, Bernardo KL, Probst KL, Woolsey TA 1992 Cortical local circuit axons do not mature after early deafferentation. Proc Natl Acad Sci USA 89:1832–1836

Melzer P, Smith CB 1998 Plasticity of cerebral metabolic whisker maps in adult mice after whisker follicle removal. I. Modifications in barrel cortex coincide with reorganization of follicular innervation. Neuroscience 83:27–41

Melzer P, Van der Loos H, Dorfl J et al 1985 A magnetic device to stimulate selected whiskers of freely moving or restrained small rodents: its application in a deoxyglucose study. Brain Res 348:229–240

Nussbaumer JC, Van der Loos H 1985 An electrophysiological and anatomical study of projections to the mouse cortical barrelfield and its surroundings. J Neurophysiol 53:686–698

Rice FL, Van der Loos H 1977 Development of the barrels and barrel field in the somatosensory cortex of the mouse. J Comp Neurol 171:545–560

Schlaggar BL, Fox K, O'Leary DDM 1993 Postsynaptic control of plasticity in developing somatosensory cortex. Nature 364:623–626

Simons DJ 1978 Response properties of vibrissa units in rat SI somatosensory neocortex. J Neurophysiol 41:798–820

Van der Loos H, Woolsey TA 1973 Somatosensory cortex: structural alterations following early injury to sense organs. Science 179:395–398

Van der Loos H, Dorfl J, Welker E 1984 Variation in pattern of mystacial vibrissae in mice. A quantitative study of ICR stock and several inbred strains. J Hered 75:326–336

Van der Loos H, Welker E, Dorfl J, Rumo G 1986 Selective breeding for variations in patterns of mystacial vibrissae of mice. Bilaterally symmetrical strains derived from ICR stock. J Hered 77:66–82

Welker C 1971 Microelectrode delineation of fine grain somatotopic organization of SmI cerebral neocortex in albino rat. Brain Res 26:259–275

Welker E, Van der Loos H 1986a Is areal extent in sensory cerebral cortex determined by peripheral innervation density? Exp Brain Res 63:650–654

Welker E, Van der Loos H 1986b Quantitative correlation between barrel-field size and the sensory innervation of the whiskerpad: a comparative study in six strains of mice bred for different patterns of mystacial vibrissae. J Neurosci 6:3355–3373

Welker E, Armstrong-James M, Van der Loos H, Kraftsik R 1993 The mode of activation of a barrel column: response properties of single units in the somatosensory cortex of the mouse upon whisker deflection. Eur J Neurosci 5:691–712 (erratum: 1993 Eur J Neurosci 5:1421)

Welker E, Armstrong-James M, Bronchti G et al 1996 Altered sensory processing in the somatosensory cortex of the mouse mutant *barrelless*. Science 271:1864–1867

Woolsey TA, Van der Loos H 1970 The structural organization of layer IV in the somatosensory region (SI) of mouse cerebral cortex. The description of a cortical field composed of discrete cytoarchitectonic units. Brain Res 17:205–242

DISCUSSION

Kaas: I was interested in the observation that the S1 area gets larger when the mice develop more whiskers. Does another type of cortex get smaller, or does the entire cortex get larger?

Welker: We observed a 10% increase in the tangential extent of the barrel cortex, which has a total area of about 2 mm^2. It will be very difficult to determine whether such a small increase at the level of S1 has an effect on the size of the total cerebral cortex. Logically, one would think that more sensory information has to be processed in other cortical areas, notably S2 and motor cortex.

Reiner: I'm interested in the mechanism of action of the cAMP knockout. cAMP is involved in the postsynaptic signalling cascade for norepinephrine and dopamine, and there is at least some literature which suggests that norepinephrine is involved in the plasticity of the development of ocular dominance columns (Kasamatsu & Pettigrew 1979, Pettigrew & Kasamatsu 1978), but is there any literature on the importance of the role of norepinephrine input in the formation of barrels?

Parnavelas: I believe that it is not known whether norepinephrine is involved in the formation of barrels.

Levitt: Patricia Gaspar and colleagues have shown that elevated levels of serotonin gives rise to major disturbances of barrel formation in the cortex, but there are no changes in the patterns that are normally seen in the thalamus and in the brainstem. There are norepinephrine changes in the same animals, but the investigators showed that the barrels can be regulated by returning serotonin levels to normal, but not norepinephrine (Vitalis et al 1998, Cases et al 1996).

Goffinet: What is the actual nature of the defect in the adenylate cyclase 1 knockout?

Welker: We do not know. Axonal development can be subdivided into the elongation phase during which the target region is reached. This phase is followed by the phase during which the axon starts to give off branches and synapses are formed. It looks like that this second phase is affected during the establishment of thalamocortical connections in the *brl* mouse.

Goffinet: The lack of competition is not apparent in the thalamus, because there the barrels are formed correctly.

Welker: At the level of the cortex, competition does not seem to play a major factor in barrel formation: pruning of the arbors of thalamocortical axons does not seem to occur (Catalano et al 1996).

Rubenstein: Is adenyl cyclase 1 just expressed in the cortex, and not in the thalamus? If so, could this be the reason why barrels are formed in the thalamus but not in the cortex? Also, if it is expressed broadly, what are some of the phenotypes, in terms of cortical regionalization of function, in these animals?

Welker: During development adenyl cyclase 1 is expressed both in thalamus and cerebral cortex (Matsuoka et al 1997). Barrels only form in barrel cortex, so in that sense it is difficult to find similar defects in other areas.

Pettigrew: Your talk nicely illustrated the role of segregation versus fusion. You have generated a subtle disturbance of segregation, and observed a dramatic change in the way those units interpret time differences. The advantages of preserving differences by segregating the two inputs and then fusing them later on are obvious in your model.

Molnár: Many other knockouts that are defective in receptor-mediated signalling molecules have recently been generated, for example GAP43 (Maier et al 1998), phospholipase Cβ1 (Hannan et al 1998), the NMDA receptor subunit 1 (Iwasato et al 1997) and monoamine oxidase A (Cases et al 1996). Eventually, these will help us to work out how these molecules are involved in the patterning of thalamocortical fibres and how the periphery-related cytoarchitectonic differentiation is imposed on developing cortical cells (see Molnár & Hannan 2000).

O'Leary: One of the things I found intriguing in the adenyl cyclase 1 knockout is that the barrel loss is specific to cortex, whereas in mice with a knockout of the NR1 subunit of the NMDA receptor, for example, it is possible to dissociate the loss of barrels in cortex from the loss of the equivalent patterning in subcortical structures because the trigeminal system is similarly affected in the brainstem, thalamus and cortex, and when the NR1 subunit is re-reintroduced by transgenic methods, an equivalent partial restoration of patterning is observed at all levels. Since cortical patterning in this system depends upon, and reflects subcortical patterning, it is not possible to determine whether NMDA receptor action is required for cortical barrel formation using this approach.

Molnár: Yes, I agree that it is important to look at the periphery-related patterns both in thalamus and cortex. If they are missing in thalamus, that itself can explain the absence of the cortical barrels. Iwasato et al (1997) rescued the NMDAR1 knockout mice by the ectopic expression of NMDAR1 transgene. These mice live for several weeks and whisker-related patterning of pre- and postsynaptic elements all along the trigeminal pathway, including the barrel cortex, is absent. In some forms of NMDA knockout mice Erzurumlu and his colleagues have recently showed that barreloids develop in the thalamus, but no barrels appear in the primary somatosensory cortex.

O'Leary: And they crossed that with a transgenic mouse to reintroduce the NR1 subunit, and different animals show different proportional reintroduction, but the finding is still the same, i.e. the amount of barrel formation in the cortex is entirely related to the amount of barrel formation in subcortical areas.

Rubenstein: Sometimes we find that these regulatory genes are expressed not only in the brain, but also in whiskers, for example, so a mutation in a single gene can sometimes affect multiple fields, some of which may be functionally related in evolution. Have you observed any other defects in these mice?

Welker: We have not seen any obvious differences elsewhere. There are no differences in the number of digits, and the animal has the same overall size as their genetic background.

Rubenstein: Have you looked for craniofacial defects?

Welker: No.

Kaas: I would like to raise the issue of activity again because sometimes there are no barrels in the cortex of mammals with whiskers, sometimes there are barrels in the thalamus and the brainstem but not in the cortex, and in star-nosed moles there are stripes for the rays of the nose in three different cortical areas. All of these observations may relate to activity. If you disrupt the correlated timing of activity, the formation of barrels is also disrupted. Activity is less precisely timed in the cortex, so it's easier to disrupt barrels there. You don't see whisker barrels in S2 of mice or rats because S2 represents another step of processing, and the timing of impulses induced is no longer precise. It appears possible to generate precision in timing in S2 because S2 in star-nosed moles has stripes. In the third cortical area of star-nosed moles, the stripes are the difficult to see, so the timing of impulses is probably less precise. Mutations in many genes could lead to changes that disrupt the timing of impulses. When you alter the timing, you start to degrade the system. You no longer restrict the growth of axons within a narrow region and they overlap because they don't having the timing information to tell them that they are inappropriately connecting.

Welker: We are looking at what time thalamocortical axons reach the cortical plate in normal and *barrelless* mice, and so far we have no data that indicate differences in arrival time. We are also just about to test whether there are any differences in activity.

Molnár: As we are now talking about activity, I would like to mention a few of our results that we performed in collaboration with Professor Keisuke Toyama,

Shuji Higashi and Thoru Kurotani in Kyoto Prefectural School of Medicine. The experimental idea is very simple. We dissected embryos from embryonic day (E) 16 onwards to determine when thalamocortical projections activate the cortex. We selectively stimulated the ventrobasal complex of the dorsal thalamus in whole forebrain slices and examined the elicited activation patterns in cortex. We obtained 400 μm thick sections, stained them with voltage-sensitive dyes (RH-482), put them into a differential image acquisition system developed by Fuji (Fujifilm HR Deltaron 1700), and in this way we recorded activation patterns within the cortex following selective stimulations of the thalamus with a bipolar electrode. From E16/17, we observed activation patterns in the subplate, and then these moved into the cortex. Even several days before birth the entire thickness of the cortical plate could be activated. After birth there is an interesting change in the spatiotemporal pattern of cortical activation, i.e. the initial diffuse activation pattern became more localized, eventually corresponding to barrels. We also observed a drastic change in the timing of this activation: initially the response is sluggish and lasts for more than 400 ms; and then it becomes more sharp and lasts for only 100–150 ms (Higashi et al 1996, Kurotani et al 1996).

We checked for the possibility of antidromic activation in two ways. (1) We did carbocyanine tracing experiments within the same slices. After the recordings, we fixed the slices and placed DiI (1,1'dioctadecyl-3,3,3',3'-tetramethylindocarbocyanine perchlorate) into the site of stimulation. We found that until around the age of postnatal day (P) 6, there were no or very few back-labelled cells in contrast to P21. (2) We also used the technique of current source density analysis to reveal possible antidromic activation. We put an extracellular electrode into the cortex and collected field potentials along equidistant points 50 μm apart along a line perpendicular to the lagas. We then took the second-order differential of these signals and calculated the current source density (Molnár et al 1996). We found that after blocking the synaptic activation with a cocktail of glutamate receptor antagonists, or using cobalt instead of calcium in the medium, one can diminish all the responses, and this would not be possible if there were substantial antidromic activation. Indeed, this is exactly what we observed in slices older than P6 or in the case of direct white matter stimulation, when there is a considerable amount of antidromic activation through the recurrent collaterals of corticothalamic and other corticofugal neurons. But, this type of antidromic activation was minimal or not present at all during embryonic and early postnatal period. These studies strongly suggest that the thalamocortical projections begin their interactions with the cortex from the moment they arrive, and therefore they are in a position to impose changes in cortical development.

References

Cases O, Vitalis T, Seif I, De Maeyer E, Sotelo C, Gaspar P 1996 Lack of barrels in the somatosensory cortex of monoamine oxidase A-deficient mice: role of a serotonin excess during the critical period. Neuron 16:297–307

Catalano SM, Robertson RT, Killackey HP 1996 Individual axon morphology and thalamocortical topography in developing rat somatosensory cortex. J Comp Neurol 367:36–53

Hannan AJ, Kind PC, Blakemore C 1998 Phospholipase $C\beta1$ expression correlates with neuronal differentiation and synaptic plasticity in rat somatosensory cortex. Neuropharmacology 37:593–605

Higashi S, Molnár Z, Kurotani T, Inokawa H, Toyama K 1996 Functional thalamocortical connections develop during embryonic period in the rat: an optical recording study. Soc Neurosci Abstr 22:1976

Iwasato T, Erzurumlu RS, Huerta PT et al 1997 NMDA receptor-dependent refinement of somatotopic maps. Neuron 19:1201–1210

Kasamatsu T, Pettigrew JD 1979 Preservation of binocularity after monocular deprivation in the striate cortex of kittens treated with 6-hydroxydopamine. J Comp Neurol 185:139–161

Kurotani T, Crair MC, Higashi S, Molnár Z, Toyama K 1996 The development of rat somatosensory (barrel) cortex visualized by optical recording. Protein Nucleic Acid Enzyme 41:758–765

Maier DL, Meiri KF, McCasland JS et al 1998 Absence of cortical barrels in GAP-43 knockout mice. Soc Neurosci Abstr 24:631

Matsuoka I, Suzuki Y, Defer N, Nakanishi H, Hanoune J 1997 Differential expression of type I, II, and V adenyl cyclase gene in the postnatal developing rat brain. J Neurochem 68:498–506

Molnár Z, Hannan AJ 2000 Development of thalamocortical projections in normal and mutant mice. In: Goffinet A, Rakic P (eds) Mouse brain development. Springer-Verlag, Berlin, in press

Molnár Z, Kurotani T, Higashi S, Blakemore C, Toyama K 1996 Development of functional thalamocortical synapses studied with current source density analysis in whole forebrain slices. Brain Res Assoc Abstr 13:53

Pettigrew JD, Kasamatsu T 1978 Local perfusion of noradrenaline maintains visual cortical plasticity. Nature 271:761–763

Vitalis T, Cases O, Callebert J et al 1998 Effects of monoamine oxidase A inhibition on barrel formation in the mouse somatosensory cortex: determination of a sensitive developmental period. J Comp Neurol 393:169–184

The relevance of visual perception to cortical evolution and development

Dale Purves, S. Mark Williams and R. Beau Lotto

Department of Neurobiology, Box 3209, Duke University Medical Center, Durham, NC 27710, USA

Abstract. The quality of brightness — perhaps the simplest visual attribute we perceive — appears to be determined probabilistically. In this empirical conception of the perception of light, the stimulus-induced activity of visual cortical neurons does not encode the retinal image or the properties of the stimulus *per se*, but associations (percepts) determined by the relative probabilities of the possible sources of the stimulus. If this theory is correct, the rationale for the prolonged postnatal construction of visual circuitry — and the evolution of this visual scheme — is to strengthen and/or create by activity-dependent feedback the empirically determined associations on which vision depends.

2000 Evolutionary developmental biology of the cerebral cortex. Wiley, Chichester (Novartis Foundation Symposium 228) p 240–258

As the brain matures, it grows substantially in size, a phenomenon that proceeds well beyond birth in humans and most other mammals (Fig. 1A). Since the generation of neurons in the primate brain is largely complete by late embryonic life (Rakic 1974, 1985), the same nerve cells populate the relatively small brains of neonates and the substantially larger brains of adults. This fact suggests — and several studies have confirmed — that nerve cells and their dendritic arborizations become larger and more complex as animals mature (Fig. 1B; e.g. Addison 1911, Conel 1939–1967, Altman 1972, Voyvodic 1987, Pomeroy et al 1990). In the visual system, the region of the primate brain that has been most thoroughly studied in this respect, only a minority of neural connections have been elaborated at birth (Bourgeois & Rakic 1993, see also Cragg 1975). Thus, the vast majority of neuronal branches and synapses are established in early postnatal life, during which time they are exquisitely sensitive to the influence of neuronal activity (reviewed in Purves 1994).

This prolonged plasticity of the developing brain — the visual system in particular — presents a mystery. Why should so much connectivity be established postnatally, particularly when this process incurs some risk (e.g. amblyopia and other developmental disorders generated by the susceptibility of the developing brain to insufficient or anomalous experience)? Here, we discuss evidence which indicates that the extraordinary postnatal proliferation of neuronal branches and synapses in the visual system serves to embellish the neural connections that enable the perception of an inherently ambiguous visual world by eliciting percepts according to the relative probabilities of the possible sources of visual stimuli.

Simultaneous brightness contrast

The misperception of luminance relationships in the presence of simultaneous contrast is a well-known psychophysical effect in which two test regions of the same luminance are seen as having different brightnesses when presented against different backgrounds. Thus, a grey patch on a dark background appears brighter than the same patch on a relatively light background (Fig. 2A). This phenomenon is generally attributed to the centre-surround receptive field organization of lower-order neurons in the primary visual pathway (Kuffler 1953, 1973, Wiesel & Hubel 1966). It has been apparent for more than a century, however, that at least some illusions of brightness do not depend exclusively on local contrast (reviewed in von Helmholtz 1910, Evans 1948, Beck 1972), as in the Wertheimer–Benary and Mach card illusions. Based on more sophisticated stimuli, a number of contemporary investigators have likewise concluded that perceptions of luminance are indeed difficult to explain on the basis of the receptive field properties of lower-order visual neurons (Gilchrist 1977, Gilchrist et al 1999, Knill & Kersten 1991, Adelson 1993, 1999).

If lateral interactions among lower-order visual neurons cannot easily account for the way we see luminances, then what does? The alternative explanation we have suggested is that the relative brightnesses in a scene are generated according to empirically determined probabilities. In this conception, it is not local contrast that causes differences in the brightness of the test patches, but what the luminance of the test patch, in relation to the luminances in the rest of the scene, has most often turned out to be (Williams et al 1998a,b). If percepts are generated as associations elicited by the statistics of what similar luminance profiles have represented in the past, then accumulated empirical information about luminance relationships determines what is seen.

How, then, can such reasoning account for the perceptions elicited by the standard presentation of simultaneous brightness contrast, in which two equiluminant test objects are presented on different backgrounds in the absence of any obvious information about illumination (see Fig. 2A)? The 'scene' in the standard presentation of the simultaneous brightness contrast stimulus in Fig. 2A is ambiguous in that it could represent either of the possibilities illustrated in

A

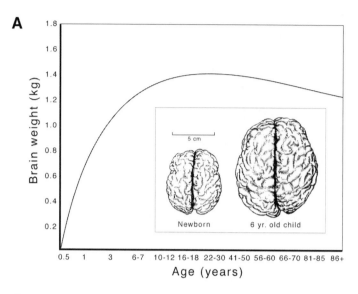

B

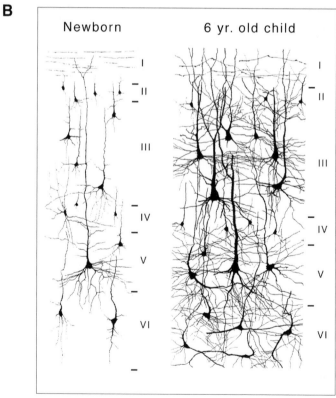

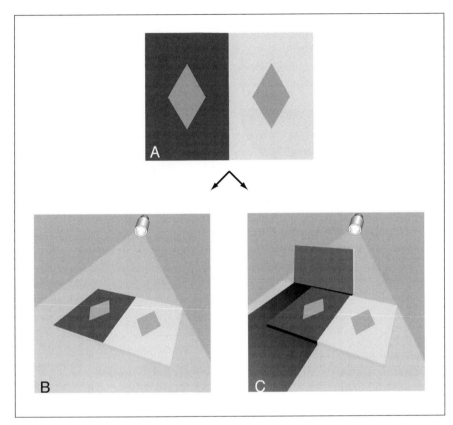

FIG. 2.   Illusion of simultaneous brightness contrast and its empirical explanation. See text for details (after Williams et al 1998b).

Fig. 2B, C (as well as many others). Additional possibilities not illustrated in Fig. 2 are a scene in which the three-dimensional shape of the object contributes to the luminance profile, a scene in which the lighter surround actually lies in shadow, and a scene in which the darker surround lies in local illumination; indeed the range of possible sources is a continuum. In short, the standard stimulus in Fig. 2A is deeply

FIG. 1.   Postnatal growth of the human brain. (A) The duration of human brain growth (according to brain weight). The growth of the brain (here based on more than 4000 neurologically normal subjects) continues for a decade or more. Inset shows the size of a normal brain at birth (left) and at age six years (right). (B) Tracings of Golgi-stained neurons in the parietal cortex of a neonatal human brain (left) and the brain of a six-year-old child (right). The marked enlargement of neurons and their increasingly complex branching during maturation implies that much postnatal brain growth arises from the ongoing elaboration of neuronal connections (A after Dekaban & Sadowski 1978, B after Conel 1939–1967).

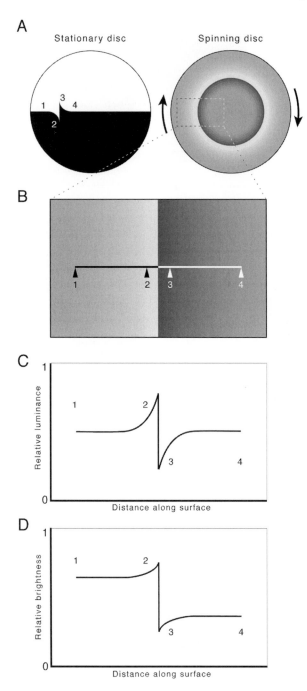

ambiguous in that it cannot uniquely signify the real-world meaning of the luminances portrayed.

An expedient — and perhaps the only — way to cope with this ambiguity (which is present in virtually all visual scenes) is to generate the perception of Fig. 2A empirically, based on what this (or any other) luminance profile has signified in the past. To our knowledge, statistical information about the meanings of luminance profiles in representative visual scenes has never been determined. Despite this ignorance, it is certain that the dark and light surrounds in Fig. 2A will not have always arisen from a surface reflectance boundary in an evenly illuminated scene (e.g. Fig. 2B), the only circumstance in which the equiluminant test diamonds would actually represent equiluminant objects. On the contrary, the stimulus will often have arisen from one of a variety of other possibilities generated by shadow, light and the spatial configuration and arrangement of the objects in the scene.

We have therefore suggested that the stimulus in Fig. 2A is seen not as the luminance profile that it actually represents, but as a statistically determined association based on synaptic linkages that have been built into the visual system by both the phylogenetic experience of the species and the ontogenetic experience of the individual. As a result, the observer perceives a construct that reflects the relative probabilities of the various possible sources of the stimulus.

Cornsweet edges

If this probabilistic theory of brightness perception has merit, then it should explain other — indeed all — the anomalous perceptions of relative light intensity that have puzzled investigators over the years. One such conundrum is the Craik–O'Brien–Cornsweet effect (Cornsweet 1970, see also Craik 1966, O'Brien 1958). In this illusion (Fig. 3A), equiluminant territories adjoining opposing light and dark luminance gradients along a step boundary are accorded

---

FIG. 3.   The Cornsweet illusion. (A) Diagram of the painted disk used by Cornsweet (1970) to demonstrate that when two equiluminant regions are separated by an edge comprising a pair of oppositely disposed luminance gradients, the adjoining territories are filled-in by illusory brightness values. (Numbers indicate corresponding points in [B] and [C]). (B) Standard presentation of the Cornsweet stimulus, which is effectively a blow-up of a portion of the rotating disk in (indicated by the box in the right-hand panel in [A]), with the curvature removed. (C) Comparison of the photometric and perceptual profiles of the stimulus in (B). Despite the equal luminances of the territories adjoining the two gradients, the territory (1) to the left of the light gradient (2) looks lighter than the territory (4) to the right of dark gradient (3). Notice that in the illusion of simultaneous contrast (Fig. 2), the dark surround makes the equiluminant target look lighter, whereas in the Cornsweet illusion the dark gradient makes the adjoining equiluminant territory look darker (from Purves et al 1999).

A. Gradations of material properties

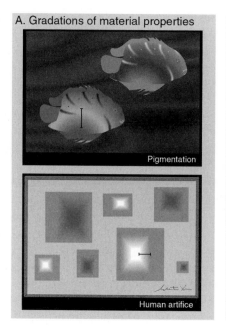

Pigmentation

Human artifice

B. Gradations of illumination

Penumbra

Curvature

FIG. 4. Sources of luminance gradients. (A) Examples of gradients arising as a result of varying reflectance properties under uniform illumination. (B) Gradients arising as a result of a varying amounts of light falling on a substrate with uniform properties: (upper panel) gradients arising from penumbras of cast shadows; (lower panel) gradients arising from the illumination of curved surfaces. Lines indicate the location and extent of the gradients in these examples.

different brightness values, thus making it obvious that the perception of the stimulus does not accord with its measured luminances. In a standard presentation (e.g. Fig. 3B), the territory adjacent to the light gradient appears brighter (lighter) than the territory adjoining the dark gradient. On average, subjects viewing this stimulus perceived a difference of 12% between the lighter and darker territories (Purves et al 1999). Thus, the Craik–O'Brien–Cornsweet illusion, like the illusions of simultaneous brightness contrast, presents equiluminant territories that are perceived differently. Indeed, the explanation of the effect suggested by Cornsweet (1970) is also based on lateral interactions of retinal or other lower-order visual neurons.

In fact, the Cornsweet stimulus is, like the standard simultaneous brightness contrast stimulus in Fig. 2, ambiguous (Fig. 4A). To understand this ambiguity and how it demands the same empirical solution, we considered the possible sources of the luminance gradients that underlie the Cornsweet edge (Fig. 4). Luminance gradients arise as a result of: (1) gradations in the material properties of an object that lead to gradual changes in surface reflectances; or (2) gradations in illumination (Fig. 4B). The territories adjoining a reflectance gradient arising from a transition in the material properties of an object (Fig. 4A) do not generally imply different amounts of light reaching the surface in question; on the contrary, such gradients and the territories adjacent to them are usually illuminated to the same degree. Conversely, gradients of illumination generally signify differences in the illumination of adjacent surfaces (see Fig. 4B). As a result, the territory flanking the lighter edge of an illumination gradient is typically more intensely lit than the territory flanking the darker edge, irrespective of whether the source of the gradient is a penumbra, a curved surface, the effect of distance from a local source or the back illumination of a semi-transparent object.

These empirical facts distinguish, in a statistical sense, the real-world behaviour of territories adjoining reflectance gradients from territories adjoining illumination gradients. Although many specific sources could underlie the standard Cornsweet stimulus (or something like it), any particular source will have represented one of these two major categories: an opposing pair of gradients arising from reflectance properties (Fig. 5A), or opposing gradients based on differences of illumination (Fig. 5B). If the source is based on reflectances, then the adjoining territories that return the same amount of light to the eye will generally have been objects with the same reflectance properties illuminated equally. Conversely, if the source of the Cornsweet edge is based on gradients of illumination, then the adjoining territories will typically have been differently reflective surfaces receiving different amounts of light. Since the stimulus *per se* lacks information that can uniquely specify it, the visual system generates an association in response to the stimulus (the percept) that incorporates these two categories of sources according to their relative probabilities. Indeed, the relative brightnesses of the territories adjoining Cornsweet edges can be greatly altered by whether the ancillary information in the scene accords with the most likely source of the adjoining equiluminant territories being similarly illuminated objects having the same material properties, or unequally illuminated objects with different material properties (Fig. 6).

Thus, like illusions of simultaneous brightness contrast, the Craik–O'Brien–Cornsweet effect can be explained in terms of a visual strategy in which percepts are generated as statistical constructs governed by the relative probabilities of the possible sources of luminance profiles in the stimulus.

Mach bands

A final phenomenon that we considered in this series of studies on the perception of luminance was Mach bands. In 1865, Ernst Mach described dark and light bands universally seen when viewing a luminance gradient that lacks any photometric basis for such percepts (Fig. 7; Mach 1865, 1959). The stimulus that Mach used

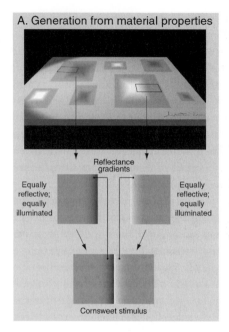

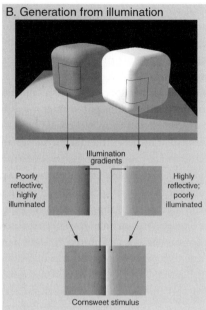

FIG. 5.   The gradients in the standard Cornsweet stimulus could arise from gradual changes in the properties of the substrate observed under the same illuminant (A), or from gradual changes in the illumination of two different substrates (B). The empirical significance of these different possible sources is that equiluminant territories adjoining luminance gradients arising from substrate qualities will typically have represented surfaces that have the same reflectance, whereas territories adjoining gradients arising from differences in illumination will typically have represented surfaces with different reflectances.

was a wheel with black and white sectors that, when spun, provided a linear gradient of luminance linking a uniformly lighter region nearer the centre of the disk with a uniformly darker region nearer its periphery. In response to this stimulus, observers perceive an illusory band of maximum lightness at the initiation of the gradient, and a band of maximum darkness at its termination. Like illusions of simultaneous brightness contrast and the Cornsweet effect, Mach bands have generally been considered a perceptual manifestation of lateral inhibitory interactions among retinal or other lower-order visual neurons (e.g. Ratliff 1965). Because this illusion bears no particular resemblance to the effects already considered, Mach bands present a special challenge to explaining the gamut of brightness experience in probabilistic terms.

    In thinking about how Mach's stimulus might elicit illusory light and dark bands in a manner akin to the empirical generation of simultaneous brightness contrast effects and the Cornsweet illusion, we again considered the real-world

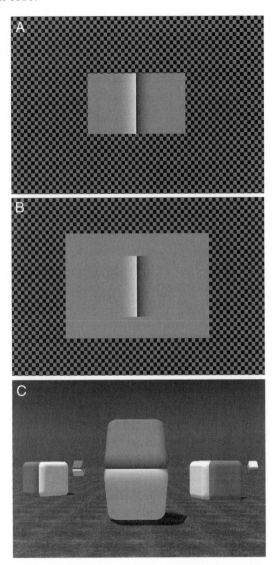

FIG. 6. Diminishment or enhancement of the Cornsweet effect by altering the relative probabilities of the possible sources illustrated in Fig. 5. (A) In the standard presentation of the Cornsweet stimulus the surface adjoining the light gradient appears lighter than the surface adjoining the dark gradient, even though the two surfaces are equiluminant. (B) When the background contrast is removed this effect is diminished: the adjoining territories are now perceived as having about the same brightness. (C) Conversely, by combining mutually reinforcing cues in a complex scene that simulates the wealth of information about the probable sources of the luminance gradients in the Cornsweet stimulus (which is the same as in panel A), the illusion can be substantially enhanced compared to the standard presentation in (A).

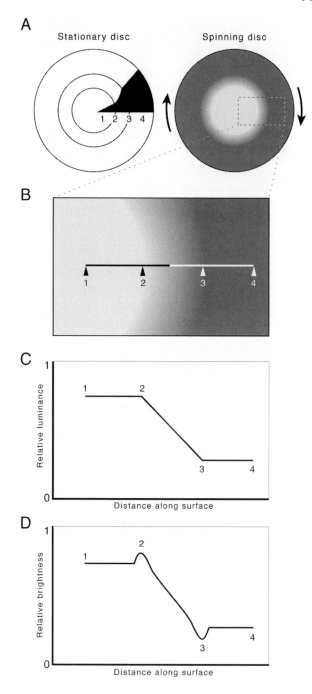

sources of the luminance gradient in the Mach stimulus (in addition to reflectance differences, penumbras and curved surfaces; see Fig. 4B) (Lotto et al 1999a,b). The typical penumbral gradient does not have photometric maxima and minima. Linear gradients generated by curved surfaces, however, typically have a highlight at their onset and a lowlight at their offset, for straightforward reasons having to do with the specular properties of most surfaces and the illumination of objects by indirect as well as direct light (Lotto et al 1999a). Thus, in considering how luminance gradients induce illusory bands of lightness and darkness, Mach and others failed to notice that the *perceptual* profile elicited by his spinning disk is remarkably similar to the overall *luminance* profile of curved surfaces.

Mach bands can therefore be understood in terms of the same probabilistic theory used to explain the standard simultaneous brightness contrast illusion and the Cornsweet effect (Fig. 8). The luminance gradient in Fig. 8A is, like the stimuli in Figs 2A and 6A, ambiguous: this standard stimulus for the Mach band illusion could represent a shadow cast on a flat surface (panel B), or a gradient preceding an attached shadow on a curved surface (panel C) (as well as a painted surface, which is the source of the stimulus in the laboratory). According to the theory, the association elicited in response to any such gradient is a construct based on the relative probabilities of the possible sources of the stimulus (determined by the observer's, and the species', past experience with similar stimuli whose significance will have been ascertained by whether or not the ensuing visually guided behaviour was successful). In the case of the stimulus in Fig. 8A, the features of this linear luminance profile — like the features of Mach's spinning disk in Fig. 7 — will often have been adorned by highlights and lowlights; consequently the stimulus triggers an association that incorporates these features according to the relative probability that the source underlying the stimulus is adorned in this way.

If this interpretation is correct, then: (1) a computer-generated depiction of an attached shadow artificially lacking photometric highlights and lowlights should

---

FIG. 7. Mach bands. (A) Diagram of the painted disk used by Mach to elicit the Mach band illusion. When the disk is spun, a luminance gradient is established between points (2) and (3), which links the uniformly lighter centre of the stimulus (1) and the uniformly darker region at its periphery (4). (B) Blow-up of a portion of the stimulus in (A), indicating the nature and position of Mach bands. A band of illusory lightness is apparent at position (2), and a band of illusory darkness at position (3). (C) Because the portion of the black sector between points (2) and (3) in (A) is a segment of an Archimedean spiral, the luminance gradient generated between the corresponding points on the spinning disk is linear, as indicated by this photometric measurement along the line in (B). (D) Diagram of the perception of the photometric profile in (C), indicating the illusory lightness maximum at the initiation of the linear gradient (2), and the illusory minimum at its termination (3) (from Lotto et al 1999a).

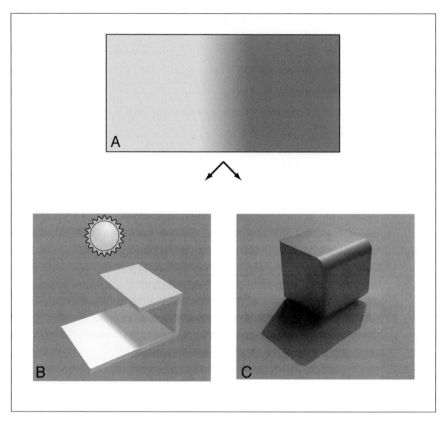

FIG. 8.   Mach bands explained in the same framework used to rationalize simultaneous brightness contrast and Cornsweet illusions. The luminance gradient in (A) is ambiguous: the profile could be the penumbra of a cast shadow (B) or the gradient generated by a curved surface (C). The penumbral gradients of cast shadows lack photometric highlights and lowlights, whereas the gradients generated by curved surfaces typically have luminance maxima and minima (after Lotto et al 1999b).

manifest Mach bands in the position that the highlight and lowlight would normally occupy; (2) the perception of Mach bands should be enhanced when the Mach stimulus is made more like the gradients arising from curved surfaces (which normally manifest highlights and lowlights), and diminished when the stimulus is made more like the penumbral gradients of cast shadows (which typically lack photometric highlights and lowlights); and (3) the salience of Mach bands in response to a given luminance gradient should be changed by ancillary cues that indicate whether the gradient pertains to a curved or a flat surface. When we tested these predictions, each of them was met (Lotto et al 1999b).

Conclusion

The perception of simultaneous brightness contrast illusions, the Craik–O'Brien–Cornsweet effect, and Mach bands can all be explained in terms of the probabilistic operation of a fundamentally empirical strategy of vision in which percepts are constructed on the basis of the relative probabilities of the possible sources of the stimulus in question. The probabilities are in turn instantiated as patterns of synaptic connections determined by the experience of the species during the course of evolution, and by the activity-dependent feedback effects of experience on neuronal growth and development. The biological advantage of this strategy is the production of visually guided behaviour that will always have the highest probability of dealing successfully with stimuli whose sources are necessarily uncertain. This strategy of vision may explain why primates have evolved such extensive visual association cortices, and why the developing visual system elaborates the vast majority of its connections postnatally under the influence of activity-dependent feedback.

Acknowledgements

This work was supported by a National Institutes of Health grant no. NS29187.

References

Addison WHF 1911 The development of the Purkinje cells and of the cortical layers in the cerebellum of the albino rat. J Comp Neurol 21:459–487

Adelson EH 1993 Perceptual organization and the judgement of brightness. Science 262:2042–2044

Adelson EH 1999 Light perception and lightness illusions. In: Gazzaniga MS (ed) The new cognitive neurosciences, 2nd edn. MIT Press, Cambridge, MA

Altman J 1972 Postnatal development of the cerebellar cortex in the rat. II. Phases in the maturation of Purkinje cells and of the molecular layer. J Comp Neurol 145:399–463

Beck J 1972 Surface color perception. Cornell University Press, Ithaca, NY

Bourgeois J-P, Rakic P 1993 Changes of synaptic density in the primary visual cortex of the macaque monkey from fetal to adult stage. J Neurosci 13:2801–2820

Conel JL 1939–1967 The postnatal development of the human cerebral cortex, vols 1–8. Harvard University Press, Cambridge, MA

Cornsweet T 1970 Visual perception. Academic Press, New York

Cragg BG 1975 The development of synapses in kitten visual cortex during visual deprivation. Exp Neurol 46:445–451

Craik KJW 1966 The nature of psychology: a selection of papers, essays and other writings by the late Kenneth JW Craik. Cambridge University Press, Cambridge

Dekaban AS, Sadowsky D 1978 Changes in brain weights during the span of human life: relation of brain weight to body heights and body weights. Ann Neurol 4:345–356

Evans RM 1948 An introduction to color. Wiley, New York

Gilchrist AL 1977 Perceived lightness depends on perceived spatial arrangement. Science 195:185–187

Gilchrist AL, Kossyfidis C, Bonato F et al 1999 An anchoring theory of lightness perception. Psychol Rev 106:795–834

Knill DC, Kersten D 1991 Apparent surface curvature affects lightness perception. Nature 351:228–230

Kuffler SW 1953 Discharge patterns and functional organization of mammalian retina. J Neurophysiol 16:37–68

Kuffler SW 1973 The single-cell approach in the visual system and the study of receptive fields. Invest Ophthalmol 12:794–813

Lotto RB, Williams SM, Purves D 1999a An empirical basis for Mach bands. Proc Natl Acad Sci USA 96:5239–5244

Lotto RB, Williams SM, Purves D 1999b Mach bands as empirically derived associations. Proc Natl Acad Sci USA 96:5245–5250

Mach E 1865 Über die Wirkung der raümlichen Verthelung des Lichtreizes auf die Netzhaut. I. Sitzungsber Kaiserlichen Akad Wiss Math-Natwiss Cl 52:303–322

Mach E 1959 The analysis of sensations and the relation of the physical to the psychical. Dover, New York

O'Brien V 1958 Contour perception, illusion and reality. J Optical Soc Am 48:112–119

Pomeroy SL, LaMantia A-S, Purves D 1990 Postnatal construction of neural circuitry in the mouse olfactory bulb. J Neurosci 10:1952–1966

Purves D 1994 Neural activity and the growth of the brain. Cambridge University Press, Cambridge

Purves D, Shimpi A, Lotto RB 1999 An empirical explanation of the Cornsweet effect. J Neurosci 19:8542–8551

Rakic P 1974 Neurons in rhesus monkey visual cortex: systematic relation between time of origin and eventual disposition. Science 183:425–427

Rakic P 1985 DNA synthesis and cell division in the adult primate brain. Ann NY Acad Sci 457:193–211

Ratliff F 1965 Mach bands: quantitative studies on neural networks in the retina. Holden-Day, San Francisco, CA

von Helmholtz H 1910 Handbuch der Physiologischen Optick. (Engl transl: Southall JPC (ed) 1962 Helmholtz's treatise on physiological optics. Dover, New York)

Voyvodic J 1987 Development and regulation of dendrites in the rat superior cervical ganglion. J Neurosci 7:904–912

Wiesel TN, Hubel DH 1966 Spatial and chromatic interactions in the lateral geniculate body of the rhesus monkey. J Neurophysiol 29:1115–1156

Williams SM, McCoy AN, Purves D 1998a The influence of depicted illumination on perceived brightness. Proc Natl Acad Sci USA 95:13296–13300

Williams SM, McCoy AN, Purves D 1998b An empirical explanation of brightness. Proc Natl Acad Sci USA 95:13301–13306

DISCUSSION

Hodos: I would like to ask you to explain the Gelb effect, i.e. if you take a black disc and shine a bright light on it in such a way that all the light falls on the disc and none falls on the surrounding area, the disc appears to be absolutely snowy white, even though you know it is black. Then, if you take a piece of white paper and put it in the path of the light, the disc appears to be black (see Osgood 1953, p 274–275).

Purves: This phenomenon can presumably be explained in the same way. Whether you see something as black or white has little to do with the photometric intensities of the stimulus components, and everything to do with their relative intensities. The perception of this relationship seems very likely to be empirical.

Hodos: How can you explain this effect when there is no obvious surround?

Purves: There is always a surround.

Hodos: Another example is the situation in which there is a black patch and a white patch, and a grey circle straddling the patches. The half of the circle on the black looks brighter than the half on the grey. This isn't surprising, but if you then take a piece of cardboard, for example, line it up with the border and begin to rotate it, you drag the darker part onto the lighter part (see Osgood 1953, p 235). How can you explain this?

Purves: Although it is possible to model simultaneous brightness contrast illusions in terms of the retinal output, the empirical explanation I have been arguing for is more consistent with the facts, as I explained, and is also more interesting! In relation to the themes of this symposium, the key question is, what has the visual cortex evolved to do? Evidently it has evolved to amass an enormous amount of empirical information to solve a tough biological problem, namely how to deal with the pervasive ambiguity of visual stimuli to maximize the observer's chances of survival. Seeing on the basis of the relative probabilities of the possible sources is a powerful strategy for solving the basic challenge that faces the visual system (i.e. the ambiguity of visual stimuli).

Kaas: I like these explanations because there has been a lot of talk about vision being an ill-posed problem that can't be solved by only knowing the information that comes from the periphery.

Purves: This kind of explanation must be generally right, simply because it's not possible to compute what the significance of ambiguous visual stimuli might be.

Kaas: These problems cannot be solved in the retina because it does not have the necessary computational power. Could you speculate whether a frog, for instance, could have these illusions, i.e. how much brain power do we need to do this?

Purves: I would assume that all vertebrates with vision see simultaneous brightness contrast phenomena.

Kaas: This suggests that complicated neural machinery is not required.

Puelles: Are amacrine cells involved?

Purves: Perhaps, but simultaneous brightness contrast is a perceptual phenomenon, and perception doesn't occur in the retina. Even the more sophisticated idea that one 'sees' the retinal output is, in my opinion, misguided.

O'Leary: Given that there are potential animal models for these percepts, if you raise an animal in conditions of uniform light stimulus, thereby altering visual experience, will it still develop the ability to see these percepts?

Purves: The answer is not known. The prediction would be that an individual's developmental experience will have an influence, but it's important to remember that experience is not limited to ontogeny, and that the experience of the species is just as important, if not more so. If you were to ask how this theory translates into neural connections and the cellular and molecular mechanisms that give rise to them, the answer would be the associative connections I refer to are formed according to all the usual cell biological and molecular rules. Remember, however, that the associations that I've discussed can be created by either the experience of the species, or the ontogenetic experience of the animal.

Reiner: If curved surfaces have highlights and lowlights, why is it advantageous, when you see an ambiguous curve that may not have highlights and lowlights in reality, to perceive that it does and that you therefore think it is a curved surface?

Purves: That's a good question. What you see in response to any given stimulus is *not* advantageous *per se*. For example, in the case of simultaneous brightness contrast two equiluminant targets are seen as being differently bright. This *is* a misperception, and in that sense a perceptual error. However, over the animal's lifetime, or evolutionary span of the species, there is an enormous benefit to a probabilistic strategy of perception: in the face of pervasive ambiguity, the animal will get it right most of the time. Thus the distortions we see are not advantageous; it's the strategy that's advantageous.

Reiner: It seems to me that when we encounter ambiguity, e.g. a curved surface without Mach bands, we are going to stumble over that ambiguous something.

Purves: My argument none the less is that this is your best bet to get along in the visual world without more major mishaps than necessary, even though you never see what's really there, and will, of course, sometimes stumble. It's just that the probability of stumbling is minimized.

Goffinet: Your work reminds me of some observations I made on chickens that had been raised that had no specific visual experience. When they were placed outside, they saw an eagle and seemed scared to death, even though they had never seen an eagle before. This suggests that some stimuli must generate a behavioural reaction without any experience.

Purves: There are a number of wonderful examples. The classic work was done by Tinbergen (1953) in herring gulls. Immediately after hatching the chicks show a fear reaction to the shadow of a predatory bird, without prior individual experience (although plenty of species experience).

Karten: These experiments were interesting because they were highly directional. When he reversed the direction of the shadow, it no longer looked like a bird of prey, but like a goose, which suggests that motion in itself provides an important clue.

Hodos: Not really. Later studies that examined the effects of the so-called 'hawk–goose' stimulus were also done under controlled laboratory conditions. In the

course, like to believe that our work contributes to the process. But sometimes we follow the opposite direction to the direction that will lift the lid on a particular problem. This, however, does not mean that we should necessarily follow the obvious leads at a particular time, but rather keep running in different directions. Uniformity could kill progress. For instance, if we look at homology, we shouldn't concentrate at the level of genes, cells or cortical areas, we should follow each of these up together. We should also not restrict ourselves to working on the obvious laboratory animals. I am convinced that some of the answers to these complicated questions will come from examining brain organization, development or gene expression patterns in rare species.

Papalopulu: One of the most exciting developments that has come out of this symposium is the coming together of different fields and the cross-fertilization of ideas. In addition, we are beginning to find out that the basic ground plan of the brain at an early developmental stage is constant across species, at least in terms of gene expression patterns. This again raises the question of how do you arrive at such different outcomes if you start out with the same basic gene expression patterns. I agree with Fredrick Bonhoeffer that small quantitative changes in both the levels and timing of expression are going to turn out to be important. We need to investigate more about how cells interpret different levels of gene product to achieve completely different outcomes, and how the timing of gene expression plays a role in this.

Krubitzer: I also agree with Fredrick Bonhoeffer. There are probably only a few small quantitative changes that generate large changes. I would like to consolidate our understanding of neocortical organization derived from comparative studies with our understanding of cortical development, particularly gene expression data, and harness information from both disciplines to make a new cortical field.

Levitt: I was struck by how much conservation there is. When Dale Purves was giving his presentation, I was thinking about the immune system, which also sets itself up, through species experience, to respond to environments that it may or may not have experienced. In other words, it is anticipating experience. This seems to be a conserved biological mechanism. Although genes are conserved as well, there are highly divergent readouts of conservation. There are genes that are highly conserved structurally, and they are complicated. We speak in terms of gene activities, but we don't understand what we mean by this because they are so complicated. I am hoping that we can get to the point of being more specific in what we mean by these biological activities, because the details will tell us what the developmental processes are about.

O'Leary: One of the important issues that we are on the verge of beginning to understand is the genetic regulation of arealization of the neocortex. It's fair to say that to date we know essentially nothing about this issue. We have identified some candidate genes, but it's time to move on and test the notion that these genes are really involved the process of arealization, which is fundamental to much of what we have discussed in this symposium. A related point is to understand better the relationship between the changes that take place in the diencephalon and those that take place in the neocortex, both across species and during development.

Herrup: I would like to see us deal with the issue of size, in terms of the control of neuron generation, because much of what we see in terms of changes in brain morphogenesis ultimately comes down to the control of neuron generation, and we know very little about it.

Boncinelli: I would like to see a systematic analysis of a sort of a scheme of covariance of gene expression patterns and biological structures. It would be interesting to choose not more than 10–20 types of developing and adult brains and study there the expression of, let us say, 50–100 developmental genes. I realize that this is a kind of brute force approach, but I would favour experimental efforts aiming at understanding in depth how changes in gene expression affect changes in structures and functions. After all, this is the meaning of biological inheritance with variations.

Pettigrew: Evolution has done lots of experiments, and what has impressed me about this symposium is that people have been willing to cross boundaries and go from genes right up to evolutionary diversity. In Australia, there is a group that started off studying learning in bees. They found a new gene in the bee by first looking at *Drosophila*, then they found it in mouse, where they found it was an embryonic lethal, so now they are looking at it in fetal marsupials. This is an example of where you can go if you are prepared to cross boundaries.

I would also like to have a greater understanding of the process of lamination. The cortex is laminated, but why is it laminated and how does it become laminated? I like André Goffinet's experiments, and I have some interesting ideas to find out more about what reelin might be doing. One way we could look at lamination, is to look at evolutionary diversity. Fish have got lots of laminar structures and they apparently have reelin, so this may be a good system to study, especially if you take advantage of what is known about the genetics of zebrafish.

Puelles: We cannot deal with evolution of the cortex without considering other portions of the telencephalon, and we cannot look at the cortex without also looking at the subpallium because there are many subtle interactions. We need to look at all these different elements, both at the genetic and cellular level.

Parnavelas: I would like to understand the role of genes in cortical area specification.

Welker: I am still thinking about what is *new* about the neocortex. Let me formulate it as a question to Harvey: if we would use some of the old stones of the ruins of the Roman Empire to construct a modern temple, would we call it a new temple?

Karten: I am committed to the questions of trying to identify: the nature of cells as components that are encoded by genes that regulate the CNS; the dynamics of how they are assembled; and how they make a different building, but still a building none the less. Animals deal with a relatively constant set of physical stimuli — in terms of photons, sound and pressure — and they organize it in amazingly similar ways. Vertebrates as a group are highly conserved, and at the same time we can't help but wonder at the beauty of diversity. Perhaps as biologists, we're always tossed between whether we want to see the similarities or whether we want to see the differences, but for the purpose of the evolution of a structure like the cortex, we perhaps narrow it down to the issue of amniotes. However, I ask the same question as you. What would we call this building, would we call it a neo-temple or archaea-temple?

Rakic: We have to be able to define 'new' before we can answer this question.

Reiner: Evolutionary neurobiology has gone through various cycles in the 20th century. It had a couple of boom times, then wound down as scientists exhausted all the methods and were left only with debates. We now seem to be in the third 'up' cycle of evolutionary neurobiology in this century. This is a particularly exciting period because we have many new tools, and the debates seem to be moving forward. There is real hope that this cycle may really nail some of these issues down.

Butler: I agree with Fredrick Bonhoeffer. It is becoming more and more obvious that small changes and perturbations can have profound effects downstream. The issue of lamination, how it comes about and what controls it, is crucial, as is the origin of the temporal neocortex. I would like to see someone identify the genes that are sufficient for lamination, and ectopically activate these genes in developing reptiles or birds, i.e. tinker with development across different amniotes.

Karten: The reason I thought this would be an interesting exercise is that I often listen to presentations at meetings, hear one sentence and find that it triggers all sorts of interesting thoughts in my mind. We have a wonderful group of people gathered here, my feeling is that it would have been a shame not to exchange our own ruminative notions that sit inside our souls and that drive our experiments afterwards.

Index of contributors

Non-participating co-authors are indicated by asterisks. Entries in bold indicate papers; other entries refer to discussion contributions.

Subject index